Neural Network Architectures

Neural Network Architectures

An Introduction

Judith E. Dayhoff

VNR VAN NOSTRAND REINHOLD
New York

To the memory of Margaret Oakley Dayhoff, pioneer in evolutionary biology.

Library of Congress Catalog Number 89-78147
ISBN 0-442-20744-1

Printed in the United States of America

Van Nostrand Reinhold
115 Fifth Avenue
New York, New York 10003

Van Nostrand Reinhold International Company Limited
11 New Fetter Lane
London EC4P 4EE, England

Van Nostrand Reinhold
480 La Trobe Street
Melbourne, Victoria 3000, Australia

Nelson Canada
1120 Birchmount Road
Scarborough, Ontario M1K 5G4, Canada

16 15 14 13 12 11 10 9 8 7 6 5 4 3 2 1

Library of Congress Cataloging-in-Publication Data

Dayhoff, Judith E.
Neural network architectures : an introduction / by Judith E. Dayhoff.
p. cm.
ISBN 0-442-20744-1
1. Computer architecture. 2. Neural computers. 3. Computer networks. I. Title.
QA76.9.A73D39 1990
006.3—dc20 89-78147
CIP

Contents

Preface

This book was written as a means of introducing both artificial and biological neural networks to a disparate audience comprising biologists and engineers as well as business professionals and others. More than an introduction, this text makes an excellent compendium for those who have already begun to experiment with neural networks. This is done by focusing on neural network paradigms, which are of equal use to beginners and professionals. Neural network paradigms are defined, studied, and explained by providing example applications and the equations that govern each network's computations.

This book also covers biological neural systems in a way that is clear and readable to non-specialists. Considerable depth is provided in the areas of biological neurons and their synapses, as these are the entities that we hope some day to emulate. The reader will be exposed to the remarkable complexity of biological neural systems as compared to artificial neural networks.

Emphasis is given to applications and examples using neural networks. An entire chapter is dedicated to backpropagation applications (Chapter 5), and another gives a general, comprehensive description of the applications of neural nets (Chapter 11). Each chapter that covers a neural network paradigm also includes applications for that paradigm.

The reader, after completing this book, will be able to understand the basic neural network paradigms and will grasp the general ideas behind neural network design. He or she will learn the steps of using an artificial neural network, will become familiar with a broad range of applications possibilities, and will be ready to begin designing experiments.

ACKNOWLEDGMENTS

I would like to thank those who helped to bring this book into existence. Appreciation goes to the readers of the manuscript: Edward Dayhoff, Dale Mugler, Wilson Or, Steve Fuld, Craig Will, Bruce Lindsey, Margaret Mortz, and Nancy Dayhoff, for their valuable feedback, insights, and encouragement, and to the artist, Phil Monti, who once again has proven that a picture is worth a thousand words.

Neural Network Architectures

1 Introduction

Neural networks provide a unique computing architecture whose potential has only begun to be tapped. Used to address problems that are intractable or cumbersome with traditional methods, these new computing architectures — inspired by the structure of the brain — are radically different from the computers that are widely used today. Neural networks are massively parallel systems that rely on dense arrangements of interconnections and surprisingly simple processors.

Artificial neural networks take their name from the networks of nerve cells in the brain. Although a great deal of biological detail is eliminated in these computing models, the artificial neural networks retain enough of the structure observed in the brain to provide insight into how biological neural processing may work. Thus these models contribute to a paramount scientific challenge — the brain understanding itself.

Neural networks provide an effective approach for a broad spectrum of applications. Neural networks excel at problems involving patterns — pattern mapping, pattern completion, and pattern classification. Neural networks may be applied to translate images into keywords, translate financial data into financial predictions, or map visual images to robotic commands. Noisy patterns — those with segments missing — may be completed with a neural network that has been trained to recall the completed patterns (for example, a neural network might input the outline of a vehicle that has been partially obscured, and produce an outline of the complete vehicle).

Possible applications for pattern classification abound: Visual images need to be classified during industrial inspections; medical images, such as magnified blood cells, need to be classified for diagnostic tests; sonar images may be input to a neural network for classification; speech recognition requires classification and identification of words and sequences of words. Even diagnostic problems, where results of tests and answers to questions are classified into appropriate diagnoses, are promising areas for neural networks. The process of building a successful neural network application is complex, but the range of possible applications is impressively broad.

Neural networks utilize a parallel processing structure that has large numbers of processors and many interconnections between them. These processors are much simpler than typical central processing units (CPUs). In a neural network each processor is linked to many of its neighbors (typically hundreds or thousands) so that there are many more interconnects than processors. The power of the neural network lies in the tremendous number of interconnections.

Neural networks are generating much interest among engineers and scientists. Artificial neural network models contribute to our understanding of biological models, provide a novel type of parallel processing that has powerful capabilities and potential for creative hardware implementations, meet the demand for fast computing hardware, and provide the potential for solving applications problems.

Neural networks excite our imagination and relentless desire to understand the self, and in addition equip us with an assemblage of unique technological tools. But what has triggered the most interest in neural networks is that models similar to biological nervous systems can actually be made to do useful computations, and, furthermore, the capabilities of the resulting systems provide an effective approach to previously unsolved problems.

In this volume we introduce a variety of different neural network architectures, illustrate their major components, and show the basic differences between neural networks and more traditional computers. Ours is a descriptive approach to neural network models and applications. Included are chapters on biological systems that describe living nerve cells, synapses, and neural assemblies. The chapters on artificial neural networks cover a broad range of architectures and example problems, many of which can be developed further to provide possibilities for realistic applications.

TRADITIONAL VERSUS NEURAL NETWORK ARCHITECTURE

Neural network architectures are strikingly different from traditional single-processor computers. Traditional Von Neumann machines have a single CPU that performs all of its computations in sequence. A typical CPU is capable of a hundred or more basic commands, including adds, subtracts, loads, and shifts, among others. The commands are executed one at a time, at successive steps of a time clock. In contrast, a neural network processing unit may do only one or, at most, a few calculations. A summation function is performed on its inputs; incremental changes are made to parameters associated with interconnections. This simple structure nevertheless provides a neural network with the capabilities to classify and recognize patterns, to perform pattern mapping, and to be useful as a computing tool.

The processing power of a neural network is measured mainly by the number of interconnection updates per second; in contrast, Von Neumann machines are benchmarked by the number of instructions that are performed per second, in sequence, by a single processor. Neural networks, during their learning phase, adjust parameters associated with the interconnections between neurons. Thus, the rate of learning is dependent on the rate of interconnection updates.

Neural network architectures depart from typical parallel processing architectures in some basic respects. First, the processors in a neural network are massively interconnected. As a result, there are more interconnections than there are processing units. In fact, the number of interconnections usually far exceeds the number of processing units. State-of-the-art parallel processing architectures typically have a smaller ratio of interconnections to processing units. In addition, parallel processing architectures tend to incorporate processing units that are comparable in complexity to those of Von Neumann machines. Neural network architectures depart from this organization scheme by containing simpler processing units, which are designed for summation of many inputs and adjustment of interconnection parameters.

BIOLOGICAL NEURAL SYSTEMS—THE ORIGINAL NEURAL NETWORKS

Neural network architectures are motivated by models of our own brains and nerve cells. Although our current knowledge of the brain is limited, we do have much detailed anatomical and physiological information. The basic anatomy of an individual nerve cell—or neuron—is known, and the most important biochemical reactions that govern its activities have been identified.

A diagram of a nerve cell typical of those in the human brain is shown in Figure 1-1. The output area of the neuron is a long, branching fiber called the axon. An impulse can be triggered by the cell, and sent along the axon branches to the ends of the fibers. The input area of the nerve cell is a set of branching fibers called dendrites. The connecting point between an axon and a dendrite is the synapse. When a series of impulses is received at the dendritic

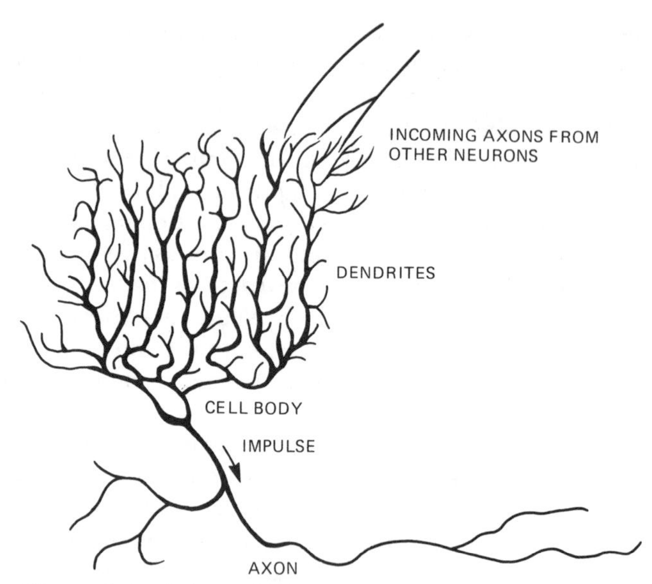

Figure 1-1. Schematic drawing of a biological nerve cell.

areas of a neuron, the result is usually an increased probability that the target neuron will fire an impulse down its axon.

The neuron shown in Figure 1-2a was photographed from a tissue culture of embryonic nerve cells. Although the axon is hidden, the dendritic tree is

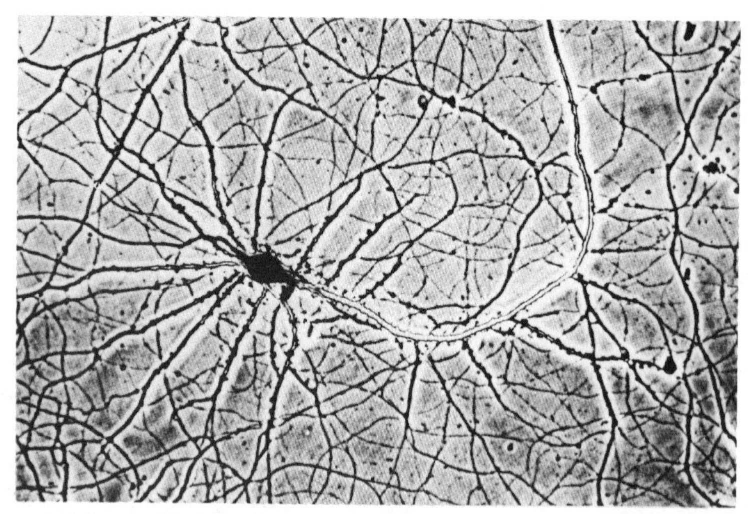

Figure 1-2a. A biological neuron magnified 400× with the dendritic tree in the foreground (courtesy of Gary Banker and Aaron Waxman, Univ. of Virginia).

apparent. The many larger fibers in the foreground are dendritic branches; the smaller fibers that crisscross in the background are axons that synapse onto the dendrites, bringing incoming impulses from other neurons.

Figure 1-2b shows a typical network of neurons, traced from the human visual cortex. These neurons appeared when a thin section of the cortex was impregnated with a Golgi stain, which is taken up by only 2% of the neurons.

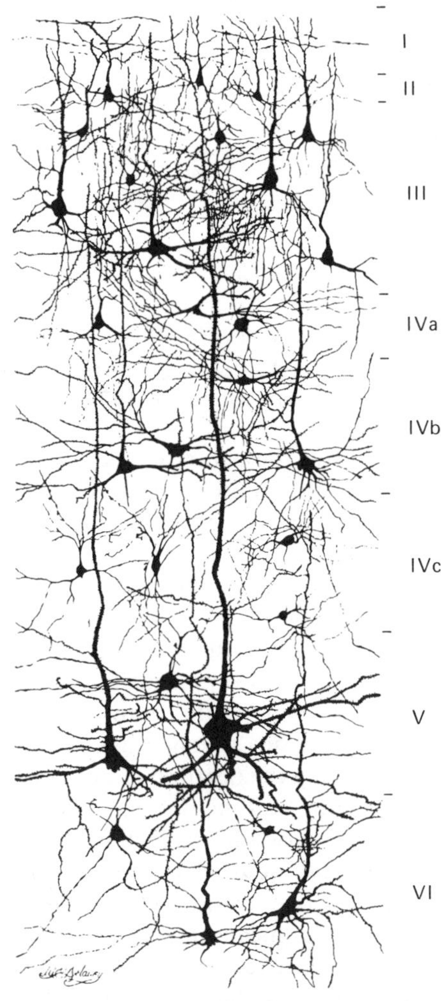

Figure 1-2b. A Golgi-stained preparation from the visual cortex of a two-year-old child showing prominent dendritic arborizations (from Conel, ***The Postnatal Development of the Human Cerebral Cortex,*** vol VI. Harvard Univ. Press, 1959).

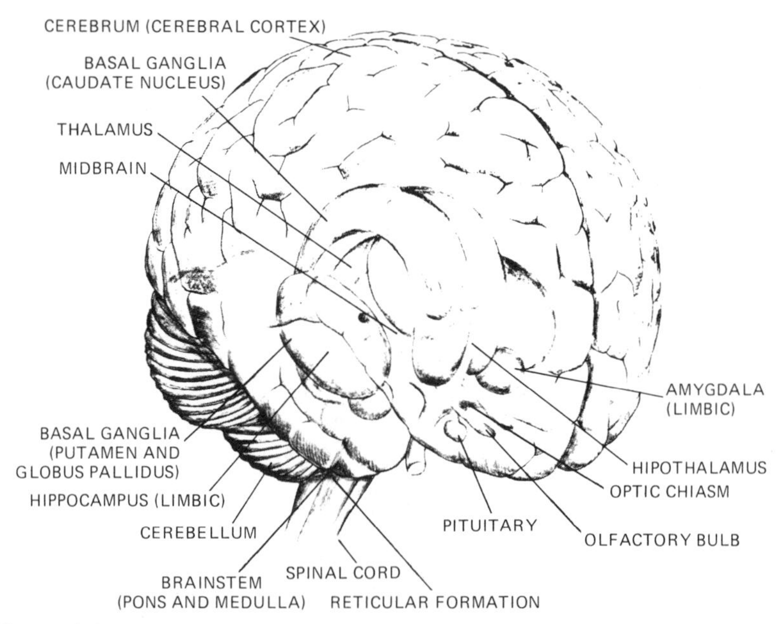

Figure 1-3. Major structures of the human brain (from Nauta and Feirtag, The organization of the brain, *Scientific American* 1979).

The resulting picture indicates the nature of the biological neural network present, with densely placed neurons and myriad intersecting nerve branches. (The actual biological network is much more dense than that shown in the figure because of the sparsity of cells that take up the Golgi stain.) This picture exemplifies the vast interconnected arrays of neurons that appear in biological neural networks.

Figure 1-3 depicts the human brain. The basic circuitry of the brain is considered in terms of general pathways. Details concerning which individual neurons are connected to which other individual neurons have not yet been mapped in the human nervous system, but considerable research effort has been put toward elucidating the detailed circuitry of the brain and determining both the fixed structure and the degree of flexibility present.

The brain is a dense neural network in which the neurons are highly interconnected. The total number of neurons in the human brain is estimated at 100 billion (DARPA Neural Network Study, 1988). Each neuron is connected to perhaps 10,000 other cells, meaning each biological neuron can send impulses that may be received by as many as 10,000 target cells.

Figure 1-4 shows a comparison of different biological nervous systems with artificial neural networks (DARPA, 1988). Speed, in terms of interconnections processed per second, is plotted against storage, measured in terms of

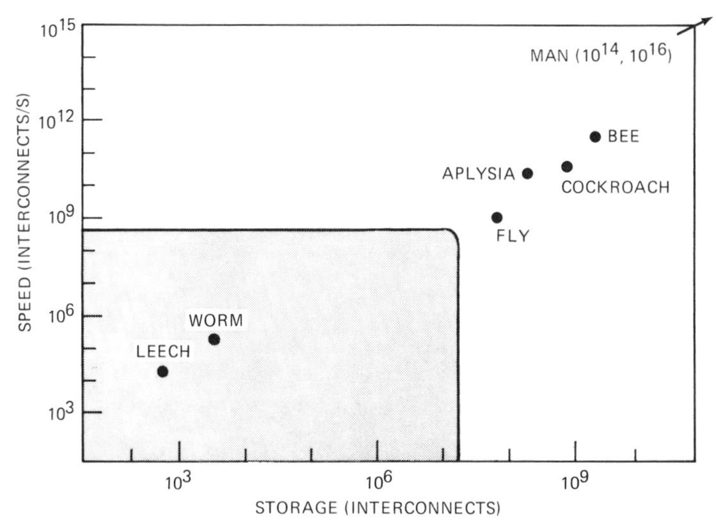

Figure 1-4. Speed versus storage for a variety of systems. Speed is measured in interconnects per second (vertical axis) and storage is measured in interconnects (horizontal axis). The shaded area shows the power of existing simulators (from DARPA Neural Network Study, 1988).

interconnections. The shaded area represents neural network sizes that are within the reach of today's artificial neural net simulations. The leech and worm, relatively primitive invertebrates, have nervous systems that appear within the range of existing simulators having fewer than 10^8 interconnections. More complex organisms, such as the fly, bee, cockroach, and aplysia (a sea slug), have nervous systems with considerably more speed and storage capacity. They appear to exceed the computational capabilities presently available in simulations. The human nervous system is far larger than the other systems plotted, and would appear beyond the top right of the graph.

ARTIFICIAL NEURAL NETWORKS—THE BASIC STRUCTURE

Figure 1-5 depicts an example of a typical processing unit for an artificial neural network. On the left are the multiple inputs to the processing unit, each arriving from another unit, which is connected to the unit shown at the center. Each interconnection has an associated connection strength, given as w_1, w_2, . . ., w_n. The processing unit performs a weighted sum on the inputs and uses a nonlinear threshold function, f, to compute its output. The calculated result is sent along the output connections to the target cells

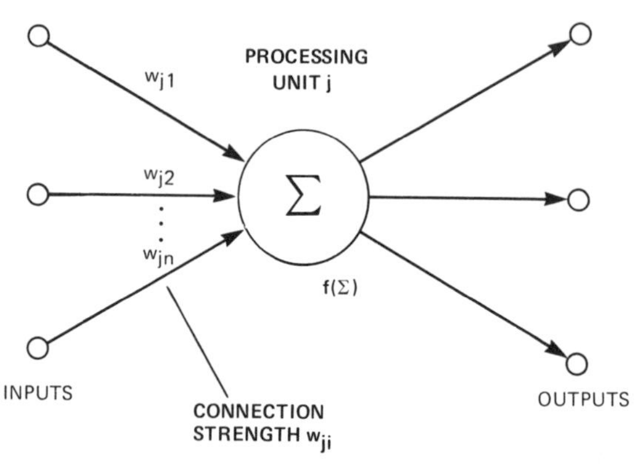

Figure 1-5. Schematic processing unit from an artificial neural network.

shown at the right. The same output value is sent along all the output connections.

The neural network shown in Fig. 1-6a has three layers of processing units, a typical organization for the neural net paradigm known as back-error propagation. First is a layer of input units. These units assume the values of a pattern, represented as a vector, that is input to the network. The middle, "hidden," layer of this network consists of "feature detectors"—units that respond to particular features that may appear in the input pattern. Sometimes there is more than one hidden layer. The last layer is the output layer. The activities of these units are read as the output of the network. In some applications, output units stand for different classifications of patterns.

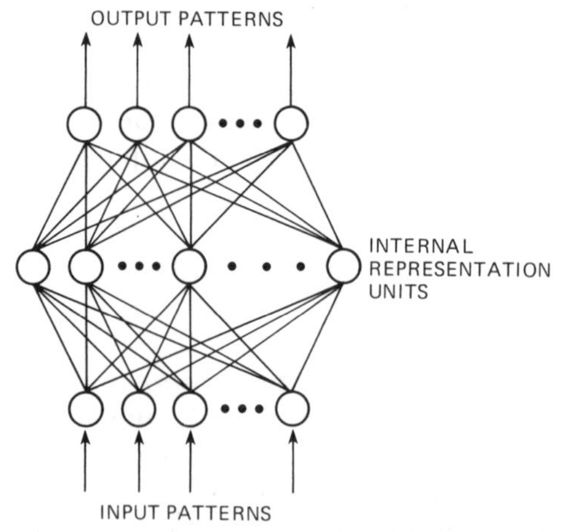

Figure 1-6a. An artificial neural network with three fully interconnected layers (from Rumelhart and McClelland, *Parallel Distributed Processing*. MIT Press, 1986).

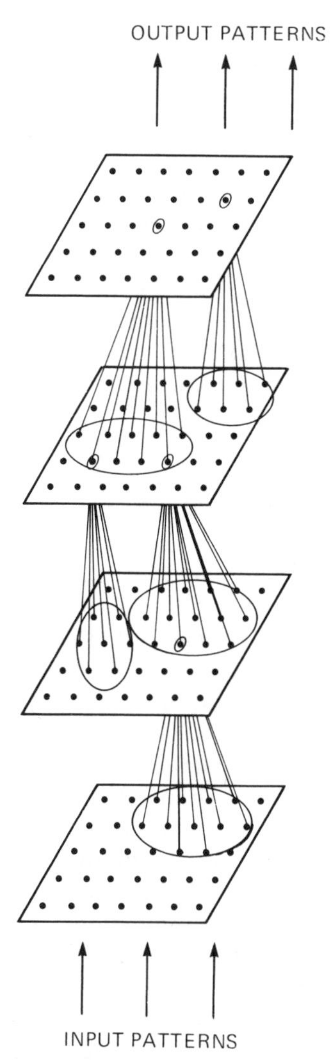

Figure 1-6b. A multilayered network with slabs of processing units that are interconnected with adjacent layers (from DARPA Neural Network Study, 1988).

A larger neural network, in which each layer is organized as a two-dimensional slab of neurons, is shown in Figure 1-6b. Neural networks are not limited to three layers, and may utilize a huge number of interconnections.

Each interconnection between processing units acts as a communication route: Numeric values are passed along these interconnections from one processing unit to another. These values are weighted by a connection strength when they are used computationally by the target processing unit. The connection strengths that are associated with each interconnection are adjusted during training to produce the final neural network.

Some neural network applications have fixed interconnection weights; these networks operate by changing activity levels of neurons without changing the weights. Most networks, however, undergo a ***training*** procedure during which the network weights are adjusted. Training may be ***supervised,*** in which case the network is presented with target answers for each pattern that is input. In some architectures, training is ***unsupervised***—the network adjusts its weights in response to input patterns without the benefit of target answers. In unsupervised learning, the network classifies the input patterns into similarity categories.

NEURAL NETWORK CHARACTERISTICS

Neural networks are not programmed; they learn by example. Typically, a neural network is presented with a training set consisting of a group of examples from which the network can learn. These examples, known as training patterns, are represented as vectors, and can be taken from such sources as images, speech signals, sensor data, robotic arm movements, financial data, and diagnosis information.

The most common training scenarios utilize supervised learning, during which the network is presented with an input pattern together with the target output for that pattern. The target output usually constitutes the correct answer, or correct classification for the input pattern. In response to these paired examples, the neural network adjusts the values of its internal weights. If training is successful, the internal parameters are then adjusted to the point where the network can produce the correct answers in response to each input pattern. Usually the set of training examples is presented many times during training to allow the network to adjust its internal parameters gradually.

Because they learn by example, neural networks have the potential for building computing systems that do not need to be programmed. This reflects a radically different approach to computing compared to traditional methods, which involve the development of computer programs. In a computer program, every step that the computer executes is specified in advance by the programmer, a process that takes time and human resources. The neural network, in contrast, begins with sample inputs and outputs, and learns to provide the correct outputs for each input.

Figures 1-7a and 1-7b contrast two different approaches to a pattern-classification problem. The task here is to classify pictures of a cat, a dog, and a

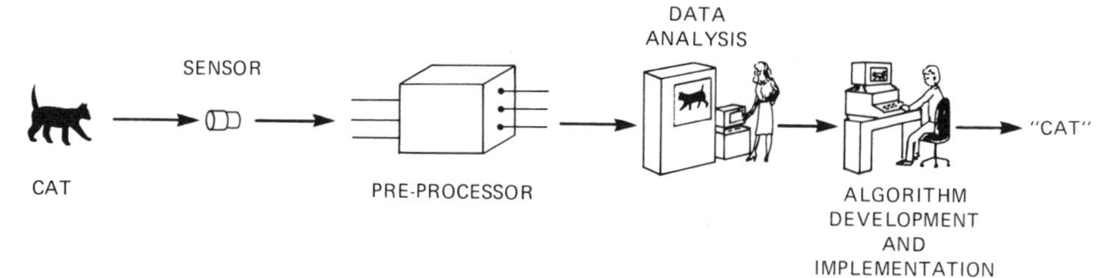

Figure 1-7a. Traditional approach to a pattern classification problem (adapted from DARPA Neural Network Study, 1988).

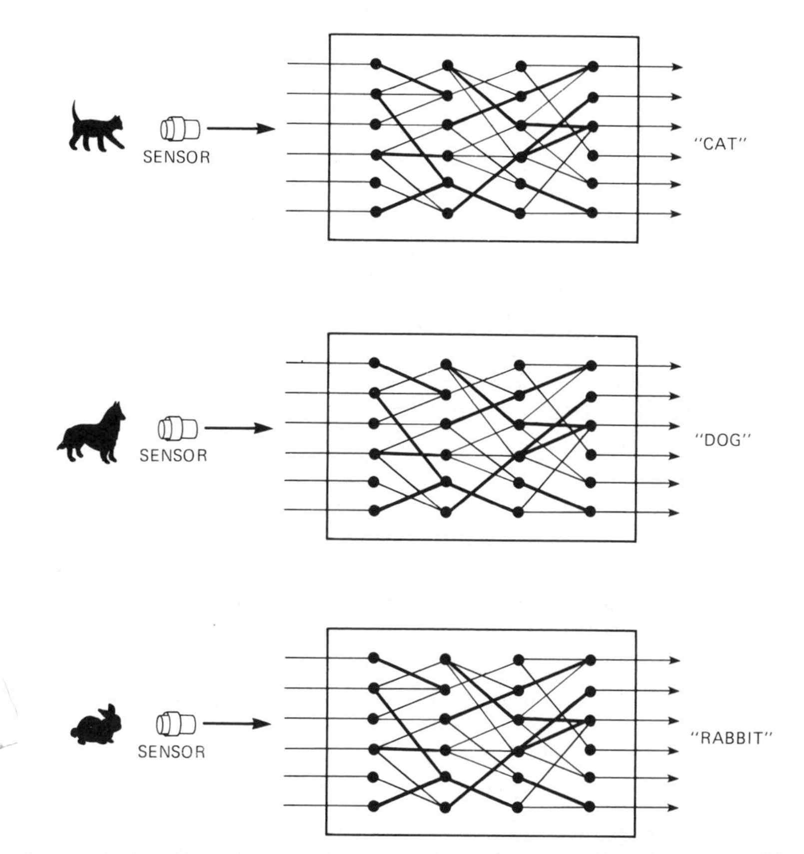

Figure 1-7b. Neural network approach to a pattern classification problem (adapted from DARPA Neural Network Study, 1988).

rabbit. Figure 1-7a illustrates the traditional approach, and is compared to a neural network approach shown in Figure 1-7b. In the traditional approach, preprocessing of the image is performed, followed by a human analysis of the data to identify important features. Human resources are then utilized in developing algorithms and programs that make use of those features to identify the cat, dog, and rabbit. The result is a program that may classify the three types of pictures, or three different programs that each recognize a single picture type. The same programs cannot then be used to classify new types of pictures.

Figure 1-7b illustrates the neural network approach. A single neural network is drawn three times in the figure. The net has already been implemented as a simulation, and may use special-purpose hardware to accelerate its computations. Preprocessing of the image data is recommended. The network is presented with the picture of the cat as an input, and with the text string "cat" as an output. The weights are readjusted automatically. The same network is then presented with the picture of a dog as an input, and "dog" as an output, and the picture of a rabbit, with "rabbit" as the output. After each presentation the weights are again readjusted automatically. This training procedure is repeated many times. After training, the same network can identify all three types of pictures. The same neural network can then be retrained to classify additional picture types, or a completely new set of pictures.

The neural network approach does not require human identification of features, or human development of algorithms and programs that are specific to the classification problem at hand, suggesting that time and human effort can be saved. There are drawbacks to the neural network approach, however: The time to train the network may not be known *a priori,* and the process of designing a network that successfully solves an applications problem may be involved. The potential of the approach, however, appears significantly better than past approaches.

Neural network architectures encode information in a distributed fashion. Typically the information that is stored in a neural net is shared by many of its processing units. This type of coding is in stark contrast to traditional memory schemes, where particular pieces of information are stored in particular locations of memory. Traditional speech recognition systems, for example, contain a lookup table of template speech patterns (individual syllables or words) that are compared one by one to spoken inputs. Such templates are stored in a specific location of the computer memory. Neural networks, in contrast, identify spoken syllables by using a number of processing units simultaneously. The internal representation is thus distributed across all or part of the network. Furthermore, more than one syllable or pattern may be stored at the same time by the same network.

Distributed storage schemes provide many advantages, the most important being that the information representation can be redundant. Thus a neural network system can undergo partial destruction of the network and may still be able to function correctly. Although redundancy can be built into other types of systems, the neural network has a natural way to organize and implement this redundancy; the result is a naturally fault- or error-tolerant system.

It is possible to develop a network that can generalize on the tasks for which it is trained, enabling the network to provide the correct answer when presented with a new input pattern that is different from the inputs in the training set. To develop a neural network that can generalize, the training set must include a variety of examples that are good preparation for the generalization task. In addition, the training session must be limited in iterations, so that no "overlearning" takes place (i.e., the learning of specific examples instead of classification criteria, which is effective and general). Thus, special considerations in constructing the training set and the training presentations must be made to permit effective generalization behavior from a neural network.

A neural network can discover the distinguishing features needed to perform a classification task. This discovery is actually a part of the network's internal self-organization. The organization of features, for example, takes place in back-propagation (described in Chapter 4). A network may be presented with a training set of pictures, along with the correct classification of these pictures into categories. The network can then find the distinguishing features between the different categories of pictures. These features can be read off from a "feature-detection" layer of neurons after the network is trained.

A neural network can be "tested" at any point during training. Thus it is possible to measure a learning curve (not unlike learning curves found in human learning sessions) for a neural network.

All of these characteristics of neural networks may be explained through the simple mathematical structure of the neural net models. Although we use broad behavioral terms such as learn, generalize, and adapt, the neural network's behavior is simple and quantifiable at each node. The computations performed in the neural net may be specified mathematically, and typically are similar to other mathematical methods already in use. Although large neural network systems may sometimes act in surprising ways, their internal mechanisms are neither mysterious nor incomprehensible.

APPLICATIONS POTENTIAL

Neural networks have far-reaching potential as building blocks in tomorrow's computational world. Already, useful applications have been designed, built, and commercialized, and much research continues in hopes of extending this success.

Neural network applications emphasize areas where they appear to offer a more appropriate approach than traditional computing has. Neural networks offer possibilities for solving problems that require pattern recognition, pattern mapping, dealing with noisy data, pattern completion, associative lookups, and systems that learn or adapt during use. Examples of specific areas where these types of problems appear include speech synthesis and recognition, image processing and analysis, sonar and seismic signal classification, and adaptive control. In addition, neural networks can perform some knowledge processing tasks, and can be used to implement associative memory. Some optimization tasks can be addressed with neural networks. The range of potential applications is impressive.

The first highly developed application was handwritten character identification. A neural network is trained on a set of handwritten characters, such as printed letters of the alphabet. The network training set then consists of the handwritten characters as inputs together with the correct identification for each character. At the completion of training, the network identifies handwritten characters in spite of the variations in the handwriting.

Another impressive applications study involved NETtalk, a neural network that learns to produce phonetic strings, which in turn specify pronunciation for written text. The input to the network in this case was English text in the form of successive letters that appear in sentences. The output of the network was phonetic notation for the proper sound to produce given the text input. The output was linked to a speech generator so that an observer could hear the network learn to speak. This network, trained by Sejnowski and Rosenberg (1987), learned to pronounce English text with a high level of accuracy.

Neural network studies have also been done for adaptive control applications. A classic implementation of a neural network control system was the broom-balancing experiment, originally done by Widrow (Widrow and Smith 1963) using a single layer of adaptive network weights. The network learned to move a cart back and forth in such a way that a broom balanced upside-down on its handle tip in the cart remained on end. More recently, applications studies were done for teaching a robotic arm how to get to its target position, and for steadying a robotic arm. Research was also done on teaching a neural network to control an autonomous vehicle using simulated, simplified vehicle control situations.

Many other applications, over a wide spectrum of fields, have been examined. Neural networks were configured to implement associative memory systems. They were applied to a variety of financial analysis problems, such as credit assessment and financial forecasting. Signal analysis has been attempted with neural networks, as well as difficult pattern-classification tasks that arise in biochemistry. In music, a string-fingering problem—that of assigning successive string and finger positions for a difficult violin passage—was studied with a neural network approach.

Neural networks are expected to complement rather than replace other technologies. Tasks that are done well by traditional computer methods need not be addressed with neural networks, but technologies that complement neural networks are far-reaching. For example, expert systems and rule-based knowledge-processing techniques are adequate for some applications, although neural networks have the ability to learn rules more flexibly. More sophisticated systems may be built in some cases from a combination of expert systems and neural networks. Sensors for visual or acoustic data may be combined in a system that includes a neural network for analysis and pattern recognition. Sound generators and speech-synthesizing electronics equipment may be combined with neural networks to provide auditory inputs and outputs. Robotics and control systems may use neural network components in the future. Simulation techniques, such as simulation languages, may be extended to include structures that allow us to simulate neural networks. Neural networks may also play a new role in the optimization of engineering designs and industrial resources.

DESIGN CHOICES

Many design choices are involved in developing a neural network application. (See Figure 1-8.) The first option is in choosing the general area of application. Usually this is an existing problem that appears amenable to solution with a neural network. Next the problem must be defined specifically so that a selection of inputs and outputs to the network may be made. Choices for inputs and outputs involve identifying the types of patterns to go into and out of the network. In addition, the researcher must design how those patterns are to represent the needed information (the representation scheme). For example, in an image-classification problem, one could input the image pixel by pixel, or one could use a preprocessing technique such as a

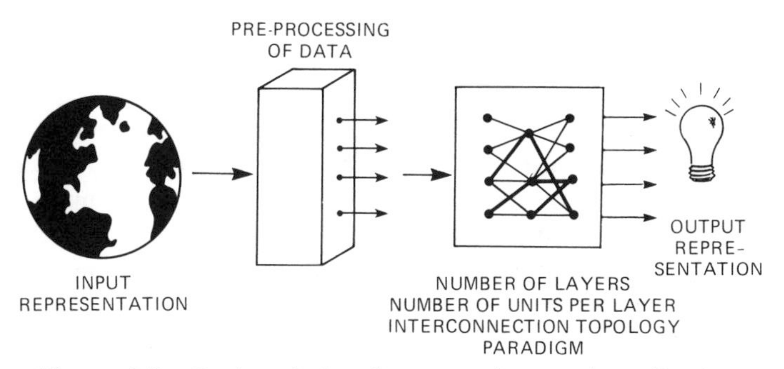

Figure 1-8. Design choices for a neural network application.

Fourier transform before the image is presented to the network. The output of the network then might have one processing unit assigned to represent each image classification, or, alternatively, a combination of several output units might represent each specific image classification.

Next, internal design choices must be made—including the topology and size of the network. The number of processing units are specified, along with the specific interconnections that the network is to have. Processing units are usually organized into distinct layers, which are either fully or partially interconnected.

There are additional choices for the dynamic activity of the processing units. A variety of neural net paradigms are available; these differ in the specifics of the processing done at each unit and in how their internal parameters are updated. Each paradigm dictates how the readjustment of parameters takes place. This readjustment results in "learning" by the network. (Paradigms covered in this volume include back-error propagation, competitive learning, Kohonen feature maps, and counterpropagation.)

Next there are internal parameters that must be "tuned" to optimize the neural net design. One such parameter is the learning rate from the back-error propagation paradigm (see Chapter 4). The value of this parameter influences the rate of learning by the network, and may possibly influence how successfully the network learns. There are experiments that indicate that learning occurs more successfully if this parameter is decreased during a learning session. Some paradigms utilize more than one parameter that must be tuned. Typically, network parameters are tuned with the help of experimental results and experience on the specific applications problem under study.

Finally, the selection of training data presented to the neural network influences whether or not the network "learns" a particular task. Like a child, how well a network will learn depends on the examples presented. A good set of examples, which illustrate the tasks to be learned well, is necessary for the desired learning to take place; a poor set of examples will result in poor learning on the part of the network. The set of training examples must also reflect the variability in the patterns that the network will encounter after training.

Although a variety of neural network paradigms have already been established, there are many variations currently being researched. Typically these variations add more complexity to gain more capabilities. Examples of additional structures under investigation include the incorporation of delay components, the use of sparse interconnections, and the inclusion of interaction between different interconnections. More than one neural net may be combined, with outputs of some networks becoming the inputs of others. Such combined systems sometimes provide improved performance and faster training times.

IMPLEMENTATIONS OF NEURAL NETWORKS

Implementations of neural networks come in many forms. The most widely used implementations of neural networks today are software simulators, computer programs that simulate the operation of the neural network. Such a simulation might be done on a Von Neumann machine. The speed of the simulation depends on the speed of the hardware upon which the simulation is executed. A variety of accelerator boards is available for individual computers to speed the computations; math coprocessors, vector processors, and other parallel processors may also be used.

Simulation is key to the development and deployment of neural network technology. With a simulator, one can establish most of the design choices in a neural network system. The choice of inputs and outputs can be tested as well as the capabilities of the particular paradigm used. Realistic training sets can be tested in simulation mode.

Implementations of neural networks are not limited to computer simulation, however. An implementation could be an individual calculating the changing parameters of the network using pencil and paper. Another implementation would be a collection of people, each one acting as a processing unit, using a hand-held calculator. Although these implementations are not fast enough to be effective for applications, they are nevertheless methods for emulating a parallel computing structure based on neural network architectures.

Because the precursors of today's neural networks were built during the same period that the digital computer was being designed, digital computer simulation was not yet available. Neural networks then were made with electrical and electronic components, including resistors and motor-driven clutches. Even though these designs appeared promising, the development of the digital computer soon dominated the field, and neural networks were developed further with the use of simulation.

One challenge to neural network applications is that they require more computational power than readily available computers have, and the tradeoffs in sizing up such a network are sometimes not apparent from a small-scale simulation. The performance of a neural network must be tested using a network the same size as that to be used in the application.

The response of an artificial neural net simulation may be accelerated through the use of specialized hardware. Such hardware may be designed using analog computing technology or a combination of analog and digital. Macroscopic electronic components may be used, or the circuits may be fabricated as semiconductor devices. Development of such specialized hardware is underway, but there are many problems yet to be solved. Such technological advances as custom logic chips and logic-enhanced memory chips are being considered for neural network implementations.

The new field of optical computing opens yet another possibility for the implementation of neural networks. Although optical computing is beyond the scope of this volume, it is important to note that the optical approach to computing has been used in neural network studies. Figure 1-9 shows results from an optical implementation of associative memory that uses a computational approach similar to neural networks.

No discussion of implementations would be complete without mention of the original neural networks—biological nervous systems. These systems provided the first implementation of neural network architectures. Developed through billions of years of evolution, they use the substances available to living systems for learning and adaptation. Many details of their learning and information processing methods are still not known. However, there is some resemblance to the way that synthetic neural networks operate, although vast differences still remain. Much of what is known about biological neurons is not included in today's computational neural networks.

These differences are so basic that to the biologist the term "neural network" is a misnomer. The synthetic neural networks do not match biological systems. They do not have the learning abilities that biological systems do, and are not capable of the same variety of behavior. Borrowing the term "neural networks" from biology is somewhat like using the term "memory" for com-

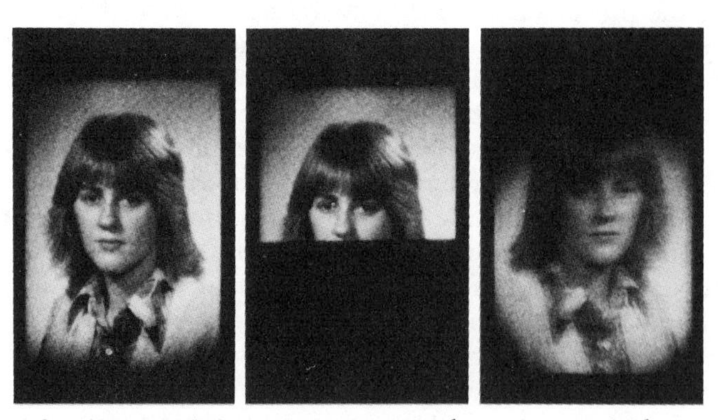

Figure 1-9. An associative memory example, using optical computing methods (from Kohonen, *Self Organization and Associative Memory,* 1988 [courtesy of Hughes Research Labs]).

puter memory, and the term "arm" for a robotic arm. In each case, the synthetic version is not the same as the original biological entity—but there is enough similarity that using the same terminology is meaningful.

IN THIS VOLUME

This book is intended to provide an overview of neural network models and architectures, and emphasizes the structure of artificial neural networks, neural network models, and the activity of biological neural systems. We present and analyze a variety of different models, so that the reader can gain a better understanding of the field taken as a whole.

We begin in Chapter 2 with a description of two early neural net paradigms: the perceptron and the adaline/madaline systems. These architectures illustrate many of the basic components of neural networks and show how supervised learning takes place. We progress to Hopfield Nets, and describe, in Chapter 3, how Hopfield Nets address optimization, associative memory, and pattern-completion tasks. Because of its importance to applications in speech, vision, knowledge processing, and pattern recognition, we then dedicate two chapters to back-error propagation. First the back-propagating architecture is covered (Chapter 4), followed by a set of simple examples that illustrate the internal organization of the network's feature detectors. Chapter 5 is devoted to back-error propagation applications, including studies of speech synthesis, speech recognition, and image processing.

Competitive layers of neurons are covered in Chapter 6, which gives illustrations of unsupervised learning and pattern classification by a simple competitive learning network. We also show a related network that performs preprocessing of noisy data with a lateral inhibition net.

Special coverage is given to biological systems and the synapses that interconnect biological neurons. In Chapter 7 we touch on topics of whole brain organization, subsystems of the brain, and nerve cell anatomy, and we show how the branching structure of axons and dendrites has an impact on the nature of interconnections between biological neurons. Chapter 8 covers biological synapses, showing the diversity of biological interconnections and their incredible complexity compared to artificial neural networks.

We continue with two chapters on additional paradigms—the Kohonen Feature Map and counterpropagation—that are particularly interesting from the standpoint of neural network applications. The dynamic activity produced by each type of network is shown, along with results of actual applications studies. Uses of these networks in continuous speech recognition and in sonar return classification are described.

In the final chapter, we review possible applications of neural networks over a broad spectrum of fields. We discuss the potential for neural networks to address such diverse problems as financial analysis, diagnosis of engine faults, diagnosis of medical symptoms, target recognition, automated industrial inspection, phoneme recognition, text-to-speech systems, sonar signal discrimination, medical image classification, compact encoding schemes, trainable robotics systems, and artificial predator-evasion systems. We describe the status of the neural network approach for each of these applications areas, and continue with a variety of issues and trade-offs that are encountered in using neural network technology. We conclude on an optimistic note for the future promise of neural networks.

References

Conel, J. L. 1959. *The postnatal development of the human cerebral cortex,* Vol. VI. Cambridge, Mass.: Harvard University Press.

DARPA Neural Network Study. 1988. Fairfax, Va.: AFCEA Press.

Kohonen, T. 1988. *Self-organization and associative memory.* New York: Springer-Verlag.

Nauta, J. W. and M. Feirtag. 1979. *Scientific American,* Sept.

Shepherd, G. M. 1988. *Neurobiology.* New York: Oxford University Press.

Widrow, B. and Smith, F. W. 1963. Pattern Recognizing Control Systems. *Computer and information sciences (COINS) symposium proceedings.* Washington, D.C.: Spartan Books.

Suggested Readings

Computer, 1988 March Vol. 21 No. 3. Los Alamitos, California: IEEE Computer Society Press.

Rumelhart, D. E., McClelland, J. L., and the PDP Research Group, 1986. *Parallel Distributed Processing,* Vol. I and II, Boston: MIT Press.

Thompson, R. F. 1985. *The Brain.* New York: W. H. Freeman.

2

Early Adaptive Networks

The capabilities of the adaptive networks of the 1950s and 1960s were limited compared to those of modern neural net architectures, but these early paradigms did bring many important properties to the attention of researchers. The two most important paradigms, the perceptron, a two-layer network designed for pattern recognition and intended as a research tool for modeling possible brain mechanisms, and the adaline/madaline system, with two or three layers of units, designed for applications studies in such areas as adaptive control, noise cancelling, and pattern analysis, were the forerunners of today's neural networks.

The perceptron and adaline/madaline systems discussed here served to introduce basic structures—including the weighted sum, weight adjustments, thresholding, and supervised learning for pattern classification—that are incorporated into today's neural networks. Even though the early paradigms had simple architectures, they had great capability in the classification of patterns. Since 1960, the complexity of neural network architectures and the size and the number of applications has increased. Some structures from the early paradigms reappear in such modern paradigms as back-error propagation, Hopfield networks, and competitive learning. In fact, both the perceptron and the adaline can be generalized to the back-error propagation system in use today.

EARLY ADAPTIVE COMPUTING

Key to exploring today's neural networks is simulation. In contrast, the first adaptive systems were built in parallel with the invention and early development of the digital computer, and thus, simulation on digital computers was not available as a means of exploring adaptive systems. The first neural network computers were based on hand-wired electronic components that produced analog computation of the adaptive nets.

The first adaptive computing machine, built by Marvin Minsky and Dean Edmonds in 1951, included automatic electric clutches, control knobs, and a gyropilot, with memory stored in the positions of the control knobs. When the machine was learning, it used the clutches to adjust its own knobs. In an interview published several years ago (Bernstein 1981), Minsky recalled:

> We sort of quit science for awhile to watch the machine. We were amazed that it could have several activities going on at once in this little nervous system. Because of the random wiring it had a sort of fail safe characteristic. If one of the neurons wasn't working, it wouldn't make much difference and with nearly three hundred tubes, and the thousands of connections we had soldered, there would usually be something wrong somewhere . . . I don't think we ever debugged our machine completely, but that didn't matter. By having this crazy random design it was almost sure to work no matter how you built it.

The capabilities of the Minsky/Edmonds machine were extremely limited. This was not the case with "the perceptron, "the first meaningful adaptive architecture, invented by Frank Rosenblatt in 1957. The perceptron was actually an entire class of architectures, composed of processing units that transmitted signals and adapted their interconnection weights. Rosenblatt's research was oriented toward modeling the brain in an attempt to understand memory, learning, and cognitive processes. As he explained (Rosenblatt 1962, 9–10):

> It is significant that the individual elements, or cells, of a [biological] nerve network have never been demonstrated to possess any specifically psychological functions, such as "memory," "awareness," or "intelligence." Such properties, therefore, presumably reside in the organization and functioning of the network as a whole, rather than in its elementary parts. In order to understand how the brain works, it thus becomes necessary to investigate the consequences of combining simple neural elements in topological organizations analogous to that of the brain. We are therefore interested in the general class of such networks, which includes the brain as a special case.

While Rosenblatt was interested in the properties of the brain, other scientists and engineers were eager to characterize the capabilities of the perceptron and to experiment with possible applications. A number of machines were built based on perceptron architectures, and hundreds of papers were written about perceptrons over the next decade. And although the research did not lead to viable applications at the time, the perceptron nevertheless

provided an architecture that was eventually extend
learning systems of today.[1]

THE PERCEPTRON

Architecture

The perceptron pattern-mapping architecture learns to classify patterns through supervised learning. The patterns it classifies are usually binary-valued (0/1) vectors, and the classification categories are expressed as binary vectors. The perceptron is limited to two layers of processing units with a single layer of adaptive weights between them. (Additional layers of weights may be added, but the additional layers are unable to adapt.)

The element shown at center of Figure 2-1 is the basic processing unit of the perceptron. Inputs arrive from the left, and each incoming interconnection has an associated weight, w_{ji}. The perceptron processing unit performs a weighted sum of its input values. The sum takes the form

$$S_j = \sum_{i=0}^{n} a_i w_{ji} \qquad (2\text{-}1)$$

where

w_{ji} = the weight associated with the connection to processing unit j from processing unit i

and

a_i is the value output by input unit i

The sum is taken over all of the units i that are inputs to the processing unit j.

The special bias input depicted at the top left of Figure 2-1 behaves as an input unit that is always fixed at the value of $+1$. Its connection to unit j has a connection weight w_{j0}, which is adjusted in the same way as all the other weights. The bias unit functions as a constant value in the sum (1) above.

---

[1] In the 1969 book *Perceptrons,* Minsky and Papert summarized and criticized (pp. 231–232) research done on this architecture. This book included negative assessments about the possibility of extending the perceptron to become a more useful computational tool and its publication led to a long dry spell in research on neural architectures for computing. We now know that extensions of the perceptron work has been fruitful for both theoretical research and applied computing in a wide variety of fields.

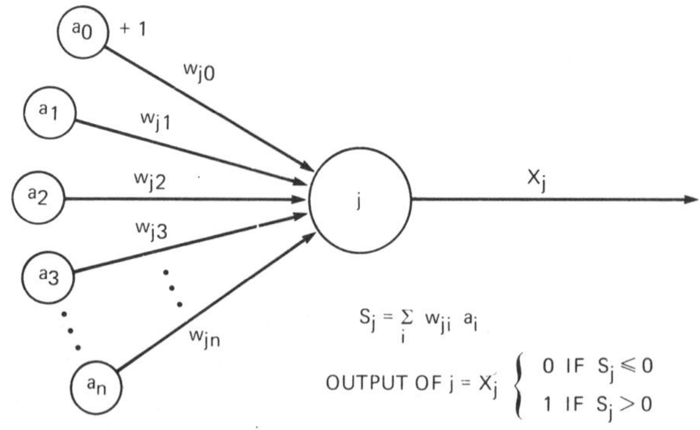

Figure 2-1. Perceptron processing unit.

The perceptron tests whether the weighted sum is above or below a threshold value, using the rule.

$$\text{if } S_j > 0 \text{ then } x_j = 1$$
$$\text{if } S_j \leq 0 \text{ then } x_j = 0 \qquad (2\text{-}2)$$

where

$$x_j = \text{the output value of processing unit } j$$

The result of (2) is the output of the perceptron processing unit, which is passed along the output line shown at the right of Fig. 2-1 and becomes an entry in the output vector for the network. (Each unit in the output layer generates such an output value.)

The two-layer perceptron network has an input layer and an output layer of processing units, as shown in Fig. 2-2. Each unit in the input layer simply uses its input value as its output. The second layer of units, then, does the computations described in (1) and (2). The two layers in Fig. 2-2 are fully interconnected, meaning every processing unit in the input layer has a connection to every unit in the output layer. There is only one layer of adjustable weights in the network; the perceptron learning rule, which corrects the weights so that the networks are more likely to produce the desired output, allows for only one layer of weights to be adjusted during learning.

A wide variety of different network topologies were used in perceptron studies during the 1950s and 1960s. Figure 2-3 shows an example of a three-layer topology with two layers of weights. In this example the inputs layer takes on the 0/1 values found in a two-dimensional grid. The grid here is a coarse image of the letter E, taken as an input. Each unit in the input layer simply uses the corresponding grid value as its output, and thus does not perform the computations described in (1) and (2).

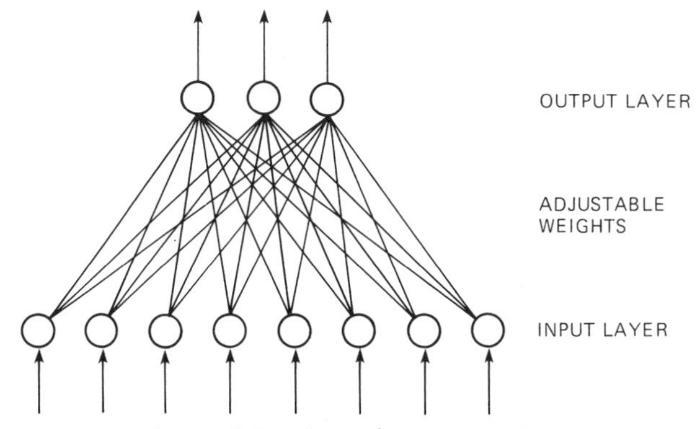

Figure 2-2. A two-layer perceptron.

The second layer of processing units displayed in Fig. 2-3 is the feature-detection layer. The incoming interconnections to this layer have randomized weights that are fixed, not adapted. The processing units in the middle layer do a summation and a threshold operation, as in (1) and (2). The feature-detection units then send signals to the output layer through a second layer of weights, which is adapted during the learning process. The layer of output units do computations (1) and (2), thereby providing the output of the network. Note that the perceptron learning rule only adapts one layer of weights (in this case, it is the second layer).

The perceptron is trained by using a training set—a set of patterns that is presented to the network repeatedly during training. Each pattern in the training set is a vector pair consisting of an input (pattern) vector, and an

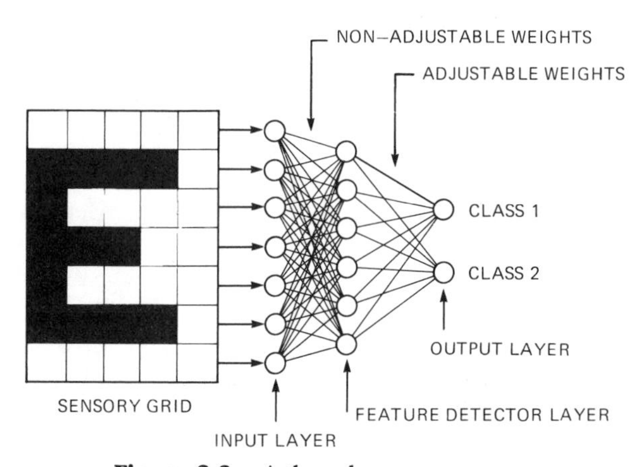

Figure 2-3. A three-layer perceptron.

output (target value) vector. Figure 2-4 shows a training set with binary vector patterns. The target outputs are shown along with each of the training patterns.

During training, each pattern in the set is presented to the network. When a pattern is input, the input layer assumes its values. The perceptron processing units then compute their outputs with the weighted sum and threshold as in (1) and (2). The network outputs are then compared to the desired outputs specified in the training set, the difference is computed, and then used to readjust the values of the connection weights. The readjustment is done in such a way that the network is—on the whole—more likely to give the desired response next time.

Figure 2-4 also shows the desired responses of the network. Each output unit is assigned to a particular classification category. The goal of the training

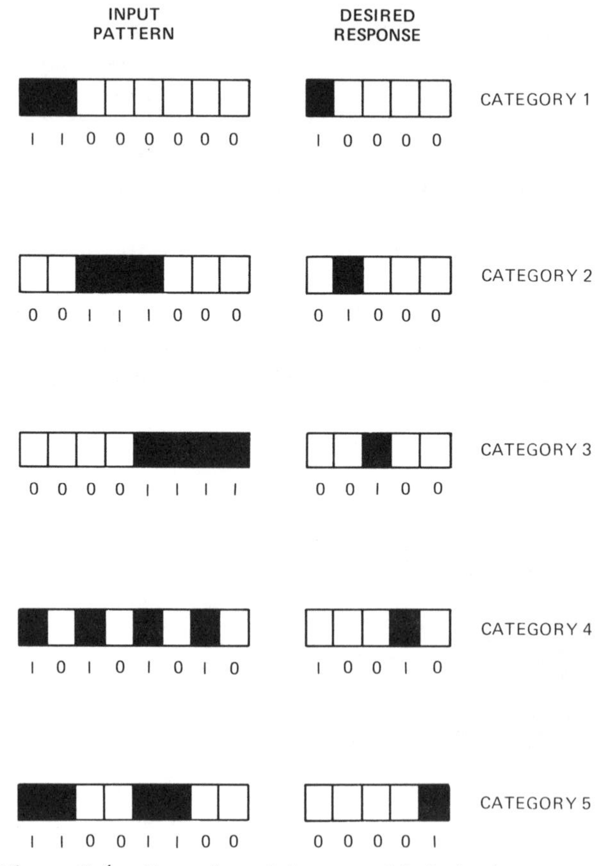

Figure 2-4. Example training set, with desired responses.

session is to arrive at a single set of weights that allow each of the mappings shown in the figure to be done successfully by the network. After training, the weights are not readjusted between the presentation of patterns.

The updating of weights has been done by a number of different rules in perceptrons (Duda & Hart 1973; Rosenblatt 1962). Using one of the simplest such rules, the difference between the target output and the network output is computed first:

$$(t_{jp} - x_{jp}) \tag{2-3}$$

where

t_{jp} = the target value for output unit j after presentation of pattern p

x_{jp} = the output value produced by output unit j after presentation of pattern p

For a perceptron that uses only 0/1 values for its units, the result of (3) is zero if the target and output are the same, and the result is +1 or −1 if they are different.

Using the simplest perceptron learning rule, a constant is added or subtracted from the appropriate weights during learning:

$$\underset{\text{new}}{w_{ji}} = \underset{\text{old}}{w_{ji}} + C\,(t_j - x_j)\,a_i \tag{2-4}$$

where

C = a small constant (the "learning rate")

$$(t_j - x_j) = \begin{cases} 1 \text{ if } t_j \text{ is 1 and } x_j \text{ is 0} \\ 0 \text{ if } t_j = x_j \\ -1 \text{ if } t_j \text{ is 0 and } x_j \text{ is 1} \end{cases}$$

a_i = 1 or 0, the value of input unit i

Rule (4) requires a weight to be changed only if its input unit is active, that is, has a value of 1. Thus, a_i must be 1 for the weight to be changed. In addition, the values of t_j and x_j must be different for the weight to be changed. If t_j and x_j are the same, then the second term in (4) is equal to 0, and the weight does not change.

According to (4), the parameter C, the "learning rate", is added to the weight when the target is higher than the output, and is subtracted from the weight when the target is lower than the output. In either case, the input unit must be active before a change is made. The value of C is usually set below 1, and determines the amount of correction made in a single iteration. The overall learning time of the network is affected by C: learning is usually slower for small values and faster for larger values of C.

Training

During training, each member of the training set is presented to the network individually, and upon each presentation the weights are readjusted. After the entire training set is presented, the set is presented again, many times. At first the performance of the network improves, but eventually performance stops improving and the network is said to have "converged." Following convergence, there are two possibilities—either the network has learned the training set successfully or it has failed to learn all of the answers correctly.

Figure 2-5 shows a sample learning curve for the problem depicted in Figure 2-4. Initially, the adaptable weights are all set to small random values, and the network does not perform very well. As the weights are adapted during training, performance improves, and when the error rate is very low, training is stopped. Performance is measured by a root-mean-squares (RMS) error value, computed by the equation

$$\sqrt{\frac{\Sigma_p \Sigma_j (t_{jp} - x_{jp})^2}{n_p n_o}} \qquad (2\text{-}5)$$

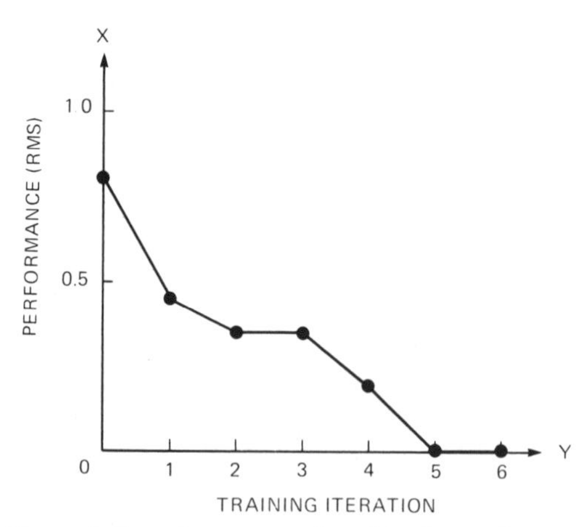

Figure 2-5. Example learning curve, showing the performance of a two-layer perceptron on the training set from Figure 2-4. $C = 0.1$ for this training session, and the initial weight values were random values between -0.05 and 0.05.

where

$$n_p = \text{number of patterns in the training set}$$

and

$$n_o = \text{number of units in the output layer}$$

The first sum is taken over all patterns in the training set, and the second sum is taken over all output processing units. The learning curve in Figure 2-5 shows that the network was trained until its RMS reached zero.

Linear Separability

A great deal of research has been done to determine what the perceptron network is actually capable of learning. Although the perceptron performs a variety of pattern classification tasks successfully, learning does not always occur, and convergence times can be extremely long. Special techniques such as adding noise sometimes help the process.

The crucial limitation of the perceptron is that it does not allow more than one layer of adaptive weights. Early researchers did not find a way of propagating the weight corrections through a multilayer network in order to make such a network learn. With more than one layer, it is not obvious what the error is for each of the hidden units and as a result, it is difficult to know how to adjust the weights so that such an error value would decrease. We now know how to extend the perceptron to include multilayered adaptive weights and such an extension has been developed as the back-error propagation paradigm (see Chapter 4).

The chief functional limitation of the perceptron is that an output unit can classify only linearly separable patterns. Figure 2-6 illustrates the general concept of linear separability—the pattern classes can be separated into two classes by drawing a single line. The figure shows four example pattern sets: the first is linearly separable and the other three are not. In this example, each pattern consists of a vector of two real numbers graphed as a single point in the diagram. Patterns in the first class are graphed as points in the black area, and patterns in the second class are graphed as points in the shaded area. The concept of linear separability may be extended to three or more dimensions, linearly separable patterns may be separated into two classes by drawing a plane or hyperplane.

The exclusive-or (Xor) function is a classic example of a pattern classification problem that is not linearly separable. The graph of the four input pairs

LINEARLY SEPARABLE REGIONS

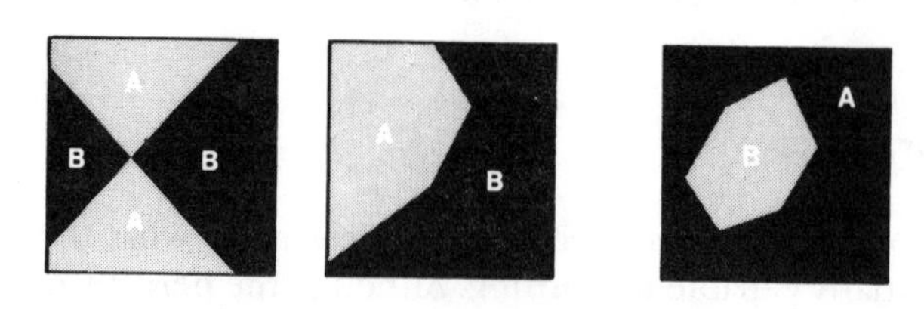

NOT LINEARLY SEPARABLE BY ONE LINE

Figure 2-6. Linear separability. The first graph is linearly separable by a single line, whereas the other three graphs are not (from DARPA Neural Network Study, 1988).

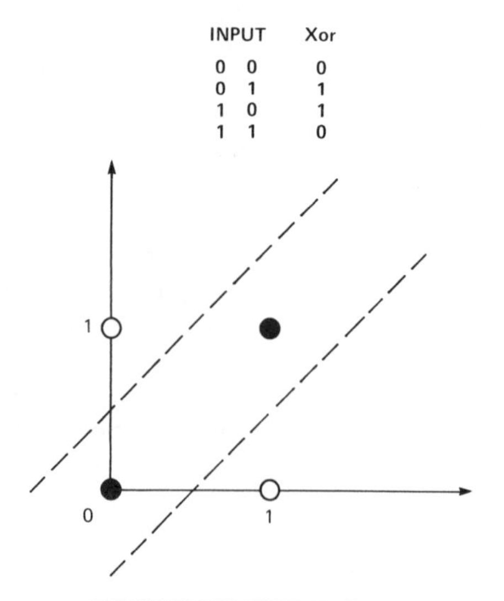

NOT LINEARLY SEPARABLE BY ONE LINE

Figure 2-7. Exclusive-or (Xor) function. Input pairs are graphed on the plane. ● = output of 0; ○ = output of 1.

(Fig. 2-7) cannot be divided by a single line to separate out the two categories (i.e., responses of 0 versus 1). (In Chapter 4 we demonstrate how the back-propagation paradigm is able to address this problem.)

ADALINES AND MADALINES

Architecture

The adaline architecture — whose name comes from *ada*ptive *li*near *ne*uron, in tribute to its resemblance to a single biological nerve cell — was invented by Bernard Widrow in 1959 and, like the perceptron, uses a threshold logic device that performs a linear summation of inputs. Its weight parameters are adapted over time.

The madaline architecture (*m*ultilayer *adaline* was also invented by Widrow and its configuration includes two or more adaline components. A number of adaline processors are taken as inputs to a madaline component, thus making a three-layer network. A number of variations of the adaline and madaline learning algorithms have been studied.

The adaline architecture is similar in design to the adaptive filters in use today (adaptive filters do adaptive noise reduction for telecommunications). A variety of other applications have been researched, including adaptive antenna arrays, adaptive blood pressure regulation, speech pattern recognition, and seismic signal pattern recognition.

Early experiments with the adaline showed that the system could only classify linearly separable patterns. Today's extensions to multilayer networks classify patterns with more complex criteria for separation. Ultimately, the adaline and madaline systems were limited by the same basic architectural constraint as the perceptron: There was no method for extending the paradigm successfully to adjusting more than one layer of weights.

Adaline Structure

Figure 2-8a shows the basic adaline unit, with the processing unit shown as a circle on the right and the input units shown on the left. A single output is present, going toward the right. Each interconnection from an input unit has an associated weight value. Figure 2-8b shows the adaline unit with its weight adjustment mechanism. The output value is compared to a target value. The difference between these two values is the error, which is used in the adaptation rule. The adaptation rule then dictates an adjustment to be made to each of the weights.

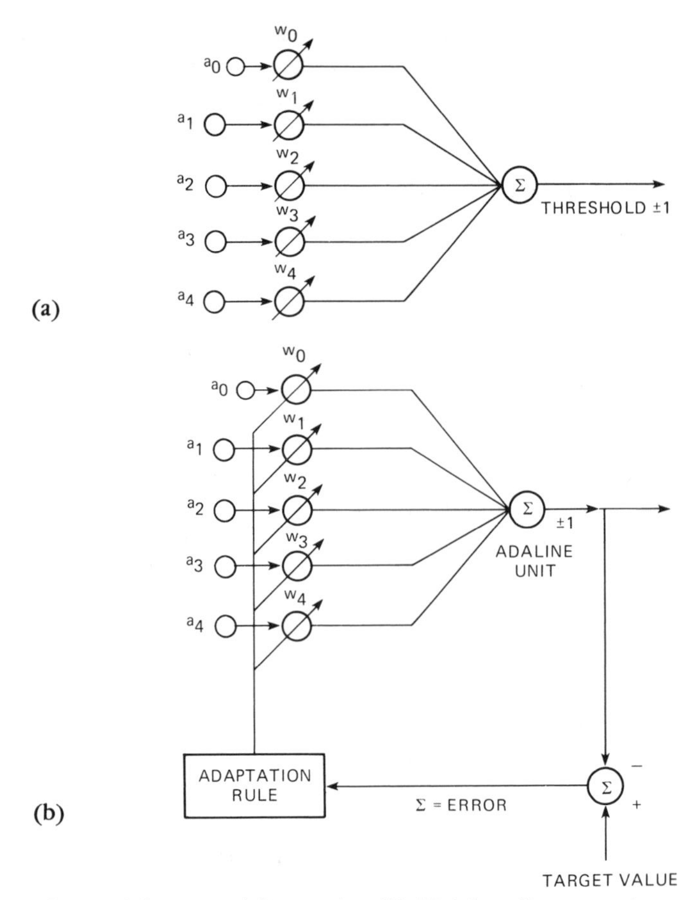

Figure 2-8. (a) Adaline unit. (b) Weight adjustment in an adaline unit. (*Source:* Adapted from Widrow et al. 1976.)

The adaline units are binary, and may take on the values of -1 or $+1$. Thus, the output value may be -1 or $+1$. These binary values may be contrasted with the perceptron architecture, which allows binary values of 0 or $+1$ only. In both cases the weights are real numbers.

The adaline unit acts as a summing device, performing a weighted sum on the inputs:

$$S = \sum_i a_i w_i \tag{2-6}$$

where

a_i = the output value of input unit i

and

$$w_i = \text{the connection weight from input } i$$

A special bias input, denoted by a_0 exists. This value is set at 1, but the connection weight from the bias unit is adjusted with the others. This unit provides a constant bias in the summation done by the adaline.

After the weighted sum is calculated as in (6), the adaline unit performs a threshold function. Generally, the threshold rule used is

$$\text{output } x_j = \begin{cases} +1 \text{ if } S \geq 0 \\ -1 \text{ if } S < 0 \end{cases} \tag{2-7}$$

Thus, the output to the adaline unit is either +1 or −1.

Training is performed on the adaline system by presenting a series of input-output pairs. The example training set in Figure 2-4 could be used (using values +1 and −1). During training, the input set is presented to the adaline, the adaline computes its output, according to (6) and (7), and the target output, taken from the training set, is then presented. During the adaptation process, the object is for the adaline to produce the target output on its own.

A number of variations exist for the adaptive learning rule of the Adaline paradigm, for example the Widrow-Hoff learning rule (Woodrow and Hoff 1960):

$$\Delta w_i = \eta \, a_i \, (t - x) \tag{2-8}$$

where η = a learning constant
a_i = output of unit i
t = target output
x = output of adaline

(The Widrow-Hoff learning rule can be extended to the learning rule in back-propagation, which is given in Chapter 4.)

Madaline Structure

The madaline system has a layer of adaline units that are connected to a single madaline unit. The madaline unit employs a majority vote rule on the outputs of the adaline layer: If more than half of the adalines output a +1, then the madaline unit outputs +1 (similarly for −1).

Figure 2-9 shows four units in the input layer, three adaline units in the second layer, and one madaline unit in the third layer. The layers are fully

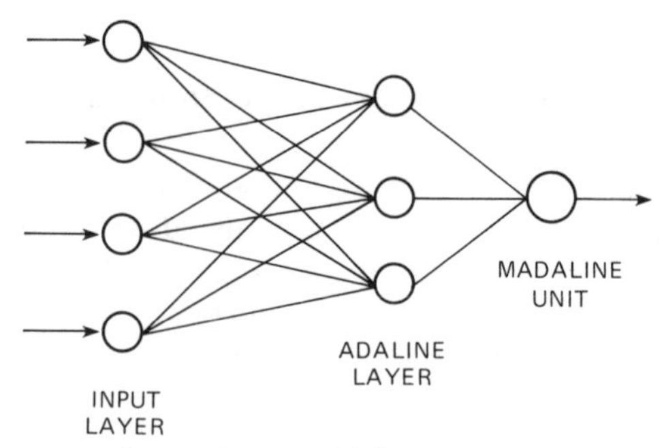

Figure 2-9. Madaline system.

interconnected, with weights associated with the interconnections from the input layer to the adaline layer. There are no connection weights associated with interconnections from the adaline layer to the madaline layer. Each adaline unit passes its output (+1 or −1) to the madaline unit, where a majority vote is taken to determine the madaline's output.

The training procedure for a madaline system is similar to that for an adaline. The training set is a set of input patterns paired with their target outputs. After the input layer is given a pattern, the madaline system computes its output, which is then compared to the target output. The weights are updated after each pattern presentation.

During the updating procedure the madaline output is first compared to the target output. Since each is +1 or −1, there is either a match, or not a match. Learning takes place only when the madaline has given an incorrect answer (i.e., not a match). Thus, if there *is* a match, nothing is changed. If there is ***not*** a match, then some of the weights are updated. Specifically, the adaline unit whose sum was closest to zero in the wrong direction is adapted. Thus, the adaptation rule assigns responsibility to the unit that can most easily assume it. Only one adaline unit updates its weights; the others keep their weights the same.

The adaline that updates may use the same adaptation rule described in (8), or a variation. Target output is taken as the target output given in the training set (± 1). The updating procedure continues until the network converges.

Applications and Experiments

When Widrow invented the adaline and madaline, his focus was on applications. He made two valuable contributions: showing what an adaline/madaline system could do and identifying candidate applications for neural networks. Widrow and his colleagues studied applications of weather-forecasting, speech recognition, vector cardiographic diagnosis, and image

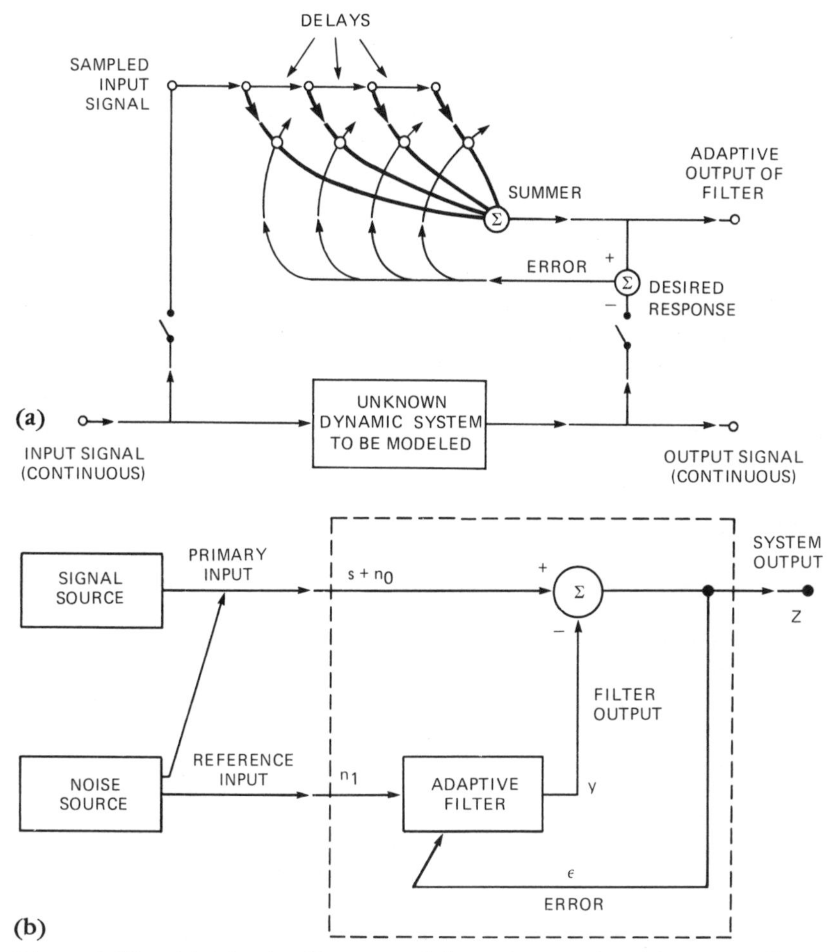

Figure 2-10. (a) Adaptive filter to model an unknown dynamic system (from Widrow and Stearn, *Adaptive Signal Processing, 1985*) (b) Adaptive noise cancelling by an adaptive filter (from Widrow et al. *Adaptive Noise Cancelling: Principles and Applications, ©Dec. 1975, IEEE*)

classification (1963–1964). They built an adaptive control system that taught a single adaline to balance a broom vertically, on end, in a moving cart, by having the adaline move the cart back and forth to keep the broom from falling (see Chapter 11).

In related research, viable applications were found for the adaptive filter. Adaptive filters are very similar to a single adaline unit, however, the adaptive filter eliminates the threshold and changes its weights continuously based on time-varying inputs and outputs.

Figure 2-10a shows the basic idea for the configuration of an adaptive filter

(Widrow et al. 1976). A signal enters at the top left, and is sampled over an interval of time to provide inputs to the processing unit (summer). The unit performs a weighted sum, compares its output to a target signal, and then weight adjustments are made. This system allows an adaptive filter to model an unknown dynamic system whose input and output signals are available.

Adaptive filter systems have been developed for noise cancelling and echo cancelling. Figure 2-10b shows the basic idea for configuring an adaptive system that does noise cancelling (Widrow & Stearns 1985). This system has a source for the signal-plus-noise and a separate channel for the noise alone. The noise channel must be from a source that is correlated with the noise. The adaptive filter element then "learns" how to transform the noise source into an appropriate value that may be subtracted from the signal to eliminate some noise. In an attempt to make the output contain only the signal and no noise, the learning feedback is set up to minimize the power of the system output.

References

Bernstein, J. 1981. Profiles: AI, Marvin Minsky. *The New Yorker,* December 14: 50–126.

Duda, R., and P. Hart. 1973. *Pattern Classification and Scene Analysis.* New York: Wiley Interscience.

Rosenblatt, F. 1962, *Principles of Neurodynamics,* Washington, D.C.: Spartan Books.

Minsky, Marvin, and S. Papert. [1969] 1988. *Perceptrons,* Expanded Edition. Cambridge, MA: MIT Press.

Widrow, B., and M. Hoff. 1960. Adaptive switching circuits. August IRE WESCON Convention Record, Part 4: 96–104.

Widrow, B., J. M. McCool, M. G. Larimore, and C. R. Johnson. 1976. Stationary and Nonstationary Learning Characteristics of the LMS Adaptive Filter. *Proceedings of IEEE.* 64(8).

Widrow, B., and S. D. Stearns. 1985. *Adaptive Signal Processing.* Englewood Cliffs, N.J.: Prentice-Hall.

3
Hopfield Networks

In a breakthrough paper published in 1982, John Hopfield introduced the network architecture that has come to be known as the Hopfield Net. In clear and simple terms he described how computational capabilities can be built from networks of neuronlike components. He illustrated an associative memory that can be implemented with his network, and later demonstrated optimization problems that could be solved. The appearance of the Hopfield network renewed interest in previous research results and precipitated a rebirth of enthusiasm for neural networks. (Fukushima, Miyake, & Ito 1983; Grossberg 1987, 1988; Reilly 1982, Anderson, 1972; Anderson et al 1977).

THE BINARY HOPFIELD NETWORK

Basic Structure

The binary Hopfield Net has a single layer of processing units. Each processing unit has an activity value, or "state" that is binary—one of two possible values. Here we use the binary states 0 and 1 (the network works the same way if values of $+1$ and -1 are used, but slight changes in the equations are required).

The entire network is considered to have a "state" at each moment. The state is a vector of 0s and 1s. Each entry in the vector corresponds to an individual processing unit in the network. Thus, at any given moment, the state of the network is represented by a state vector such as:

$$\mathbf{U} = (u_1, u_2, \ . \ . \ . \ , u_n) = (++\cdots\cdot+\cdots+)$$

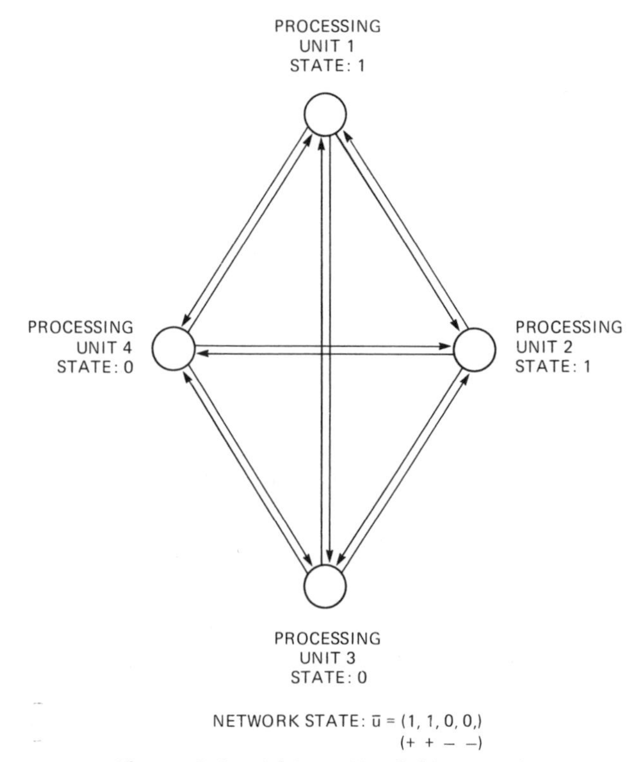

Figure 3-1. A binary Hopfield network.

This vector reflects a network of *n* processing units, where element *i* has state u_i. In this notation, a + represents a processing unit with the binary value 1, and a − represents a processing unit with the value 0. Figure 3-1 shows a diagram of the processing units in a Hopfield network, together with an example state. The state of the network can change over time as the values of individual units change.

The processing units in the Hopfield network are *fully interconnected*—each unit is connected to every other unit. In fact, the connections are "directed," and every pair of processing units has a connection in each direction (see Figure 3-2). This interconnection topology makes the network "recursive" because the outputs of each unit feed into inputs of other units in the same layer. As we shall see, this recursive organization will allow the network to relax into a stable state in the absence of external input.

Each interconnection has an associated weight. This weight is a scalar value, considered intuitively to be the *connection strength*. We let T_{ji} denote

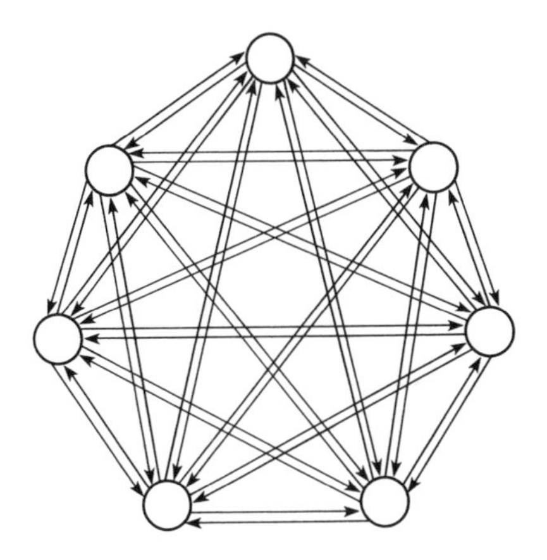

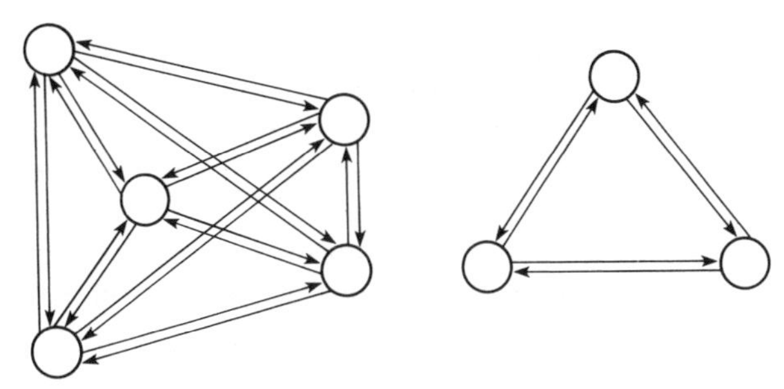

Figure 3-2. Fully interconnected one-layer networks, with connections in both directions between each pair of processing units.

the weight to unit *j* from unit *i*. In the Hopfield network, the weights T_{ji} and T_{ij} have the same value, therefore

$$T_{ji} = T_{ij}$$

Mathematical analysis has shown that when this equality is true, the network is able to converge—that is, it eventually attains a stable state. Convergence of the network is necessary in order for it to perform useful computational tasks such as optimization and associative memory. Many networks with unequal weights ($T_{ji} \neq T_{ij}$) also converge successfully.

Figure 3-3 shows an alternative method of diagramming the interconnections: The processing units appear in a row and the interconnections form a

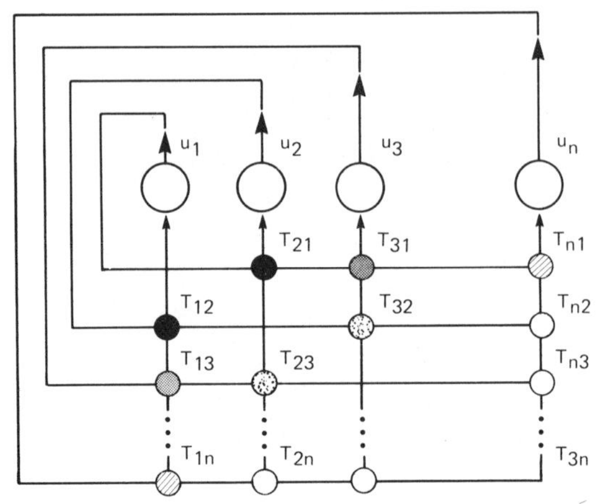

Figure 3-3. A Hopfield network, with weight values shown as circles. Shading of the circles represents the values of the weights.

gridwork. Note that with n units, there are $n(n\text{-}1)$ interconnections. Each connection weight is shown at an intersection point in the grid. The size of the weight is depicted by the shade of the circle drawn at that intersection. The weight from node i to j is shown below node j, where the line below node j intersects the line that originates at node i.

Connection weights are set at the beginning of an application. The method of setting the weights depends on the application. Two different applications will be described: the associative memory and the traveling salesman problem.

The Updating Procedure

Initially, the network is assigned a state for each processing unit. An updating procedure is applied to the units in the Hopfield network. One unit at a time is updated. The updating procedure affects the state of each unit, sometimes changing it and sometimes leaving it the same. The updating of processing units continues until no more changes can be made.

Figure 3-4 illustrates the basic processing done by a binary Hopfield Net unit during the updating procedure. Each neuron takes a weighted sum of its inputs, according to the following equation:

$$S_j = \sum_{\substack{i=1 \\ i \neq j}}^{n} u_i T_{ji} \tag{3-1}$$

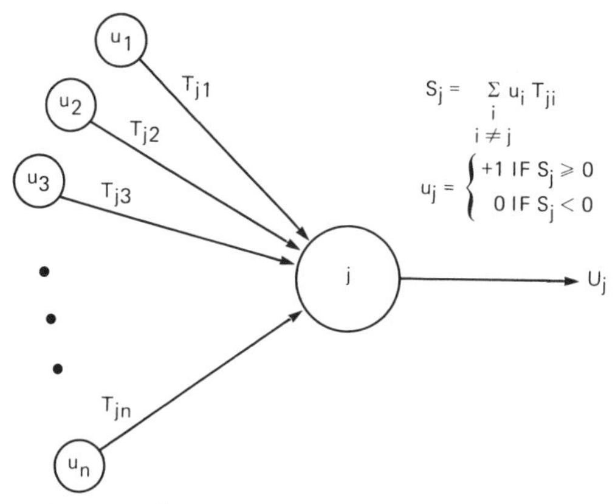

Figure 3-4. Binary Hopfield processing unit.

When this sum is calculated, the processing unit then evaluates whether the sum is greater or smaller than 0. If the sum is at least 0, then the output of the unit is set at +1. If the sum is less than 0, the output of the unit is set at 0. In mathematical terms,

$$\begin{array}{ll} \text{if } S_j \geq 0 & \text{then } u_j = 1 \\ \text{if } S_j < 0 & \text{then } u_j = 0 \end{array} \qquad (3\text{-}2)$$

Suppose that processing unit j is to be updated, as shown in Figure 3-5. The weighted sum of inputs is calculated for element j, according to Eq. (3-1). Then the rule in (3-2) is applied. The result is that unit j has the value of 0 (top of Figure 3-5). In the middle frame of Figure 3-5, another processing unit (top left) is updated, and changes its state from 1 to 0. Next, unit j is updated (bottom of Figure 3-5). The weighted sum of its inputs is now 0.5, which is positive and hence changes the state of unit j to 1.

The previous value of a processing unit is not taken into account when it is updated. The unit may or may not change its value due to the updating procedure. The example in Figure 3-5 shows a processing element that changed values through the updating procedure. Its change was caused by a prior change in the state of the unit at the top left.

One system for the updating process is to update the units in sequence, then repeat the sequence until a stable state is attained. The update mechanism posed by Hopfield is not as simple — it chooses the next unit to be updated at random, which allows all units to have the same average update rate. Eventually the network reaches a stable state: All units retain the same value upon updating. Random updating has advantages both in implementation (each unit can generate its next update time) and in function (sequential updating

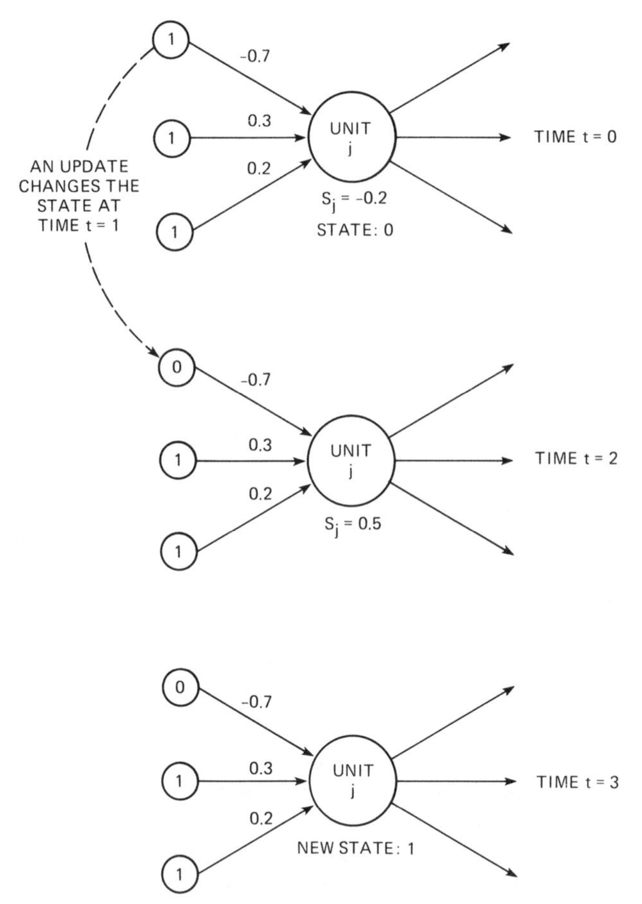

Figure 3-5. State of processing unit j at three successive times ($t = 0,1,2,3$). A change in state of the unit at top left results in a change in state for unit j.

can restrict the output states of the network in cases in which different stable states are equiprobable).

Usually, many updates must be done to all of the processing units before the network reaches a stable state. Hopfield (1982) showed that his network does eventually reach such a stable state.

The randomized updates used in the Hopfield network provide an important difference between it and other paradigms. Most other neural network paradigms have a layer of processing units updated at the same time (or nearly the same time). In contrast, the asynchronous updating of the Hopfield Net is a closer match to biological reality—biological neurons update their own states due to events that impinge upon the neuron. These impinging events are not synchronized from neuron to neuron.

Convergence

Each state of the Hopfield network has an associated "energy" value. This value is defined by:

$$E = -\frac{1}{2} \sum_{j} \sum_{\substack{i \\ j \neq j}} T_{ji} \, u_j \, u_i \tag{3-3}$$

The equation is referred to as "energy," although it does not represent the real energy of any physical system. The energy function in Eq. (3-3) is an objective function that is minimized by the network.

The successive updating of the Hopfield network provides a "convergence" procedure whereby the energy of the overall network gets smaller and smaller. Eventually the network goes into a stable state; at this stable state, the energy is at a minimum. This minimum may be local or global.

It is possible to prove that each time a processing unit is updated, the energy of the network either stays the same or decreases. As a result, this updating procedure will always allow the energy of the network to converge to a minimum.

There is also an argument that the updating procedure either decreases the energy or leaves it the same. Suppose that unit j is the next processing unit to be updated. Then, the portion of E affected by processing unit j is given by:

$$E_j = -\frac{1}{2} \sum_{\substack{i \\ i \neq j}} T_{ji} \, u_j \, u_i \tag{3-4}$$

which rearranges to

$$E_j = -\frac{1}{2} u_j \sum_{\substack{i \\ i \neq j}} T_{ji} \, u_i \tag{3-5}$$

When unit j is updated, if there is no change in its state, then the energy E_j remains the same. If there is a change in its state, then the difference in E_j is:

$$\Delta E_j = E_{j_{new}} - E_{j_{old}} = -\frac{1}{2} \Delta u_j \sum T_{ji} \, u_i \tag{3-6}$$

where

$$\Delta u_j = u_{j_{new}} - u_{j_{old}}$$

If u_j changes from 0 to 1, then

$$\Delta u_j = 1$$

and

$$\sum_i T_{ji} u_i \geq 0$$

after updating, according to Eq. (3-2).

Plugging these nonnegative values into Eq. (3-6), we get

$$\Delta E_j \leq 0$$

If u_j changes from 1 to 0, then

$$\Delta u_j = -1$$

and

$$\sum_i T_{ji} u_i < 0$$

after updating, by (3-2). Plugging these two negative values into (3-6), we get

$$\Delta E_j < 0$$

Since ΔE_j is the product of three negative numbers. Thus, the change in E is always negative or 0 no matter what change there is in the state of unit j upon updating. The network is guaranteed to converge, with E taking on lower and lower values until the network reaches a steady state.

Seeking a minimum of the energy function is analogous to seeking a minimum in a mountainous terrain. Figure 3-6 depicts a two-dimensional version of such terrain. The energy function value is reflected in the height of the graph. Each position on the terrain corresponds to a possible state of the network, and the network moves toward a minimum position in this graph. Two local minima and a global minimum are depicted. If the state of the network is changed, then a corresponding change is made in the x coordinate position on the graph. This change in turn results in a movement downhill, toward one of the minima.

The initial state of the network may be thought of as the position of a skier who has been dropped randomly onto mountainous terrain. The updating procedure moves the skier downhill until he gets to the bottom of the most accessible valley. (This valley may be at a local or a global minimum.)

Figure 3-6. An energy terrain plotted in two dimensions.

Although we may visualize this search in two or three dimensions, there are actually as many dimensions as there are processing units. The Hopfield convergence procedure seeks a minimum in such a multidimensional mountainous terrain.

In the Hopfield Net, there is no way to reach the global minimum from a local minimum. A different network paradigm, such as the Boltzmann Machine, must be used. The Boltzmann Machine uses noise to "shake" the network state out of a local minimum (Hinton & Sejnowski 1986). The Hopfield Net, however, can be restarted at a different initial position, which may allow the network to then find a global minimum.

Associative Memory

Hopfield originally proposed the application of associative memory for his binary network. This application illustrates the general approach of using a neural network architecture for developing an associative memory system. Although the binary Hopfield Net implementation is not a very efficient or reliable way to build an associative memory, the methodology is highly informative. The associative memory application illustrates the basic application technique for Hopfield networks—the energy equation is devised to model the applications problem and then used to assign the weight values. The network then converges to a solution.

The basic scheme for associative memory is shown in Figure 3-7, where each "memory" is represented by a vector of +/−'s. In this example, each memory is a person's name, together with a color associated with the name. Each memory vector is a state of the network that corresponds to a minimum in the terrain defined by the network energy. When the network starts at an initial state, the updating procedure moves the state of the network until it is at a minimum. That minimum, then, is expected to correspond to one of the "memories" of the network. Thus the network may "converge" to the stored memory that is most similar or most accessible to the initial state.

Figure 3-7a shows three example memories that may be stored in a Hopfield network with 24 processing units. In this scheme, (--++++----+-) stands for Karen. If the network is presented with this code for Karen, followed by a random string of -'s instead of a color association, then the network is expected to converge to the representation for Karen-Green. In this sense, the network has stored the association of Karen with Green (Figure 3-7b).

A similar use for the associative memory, that of recall from a noisy pattern pair — is shown in Figure 3-7c. A noisy version of the pattern pair Paul-Blue is input to the network. The network is then expected to converge to the correct +/− code for Paul-Blue.

Figure 3-8 illustrates another candidate application of the associative memory. An array of black/white pixels (Figure 3-8a) is used to represent individual characters. (Figure 3-8). Each of these characters can be represented as a +/− vector, and thus may be a memory in a Hopfield network. Associative recall allows the possibility that some fraction of the array entries be correct initially, with the rest incorrect or random (noisy). The network then begins

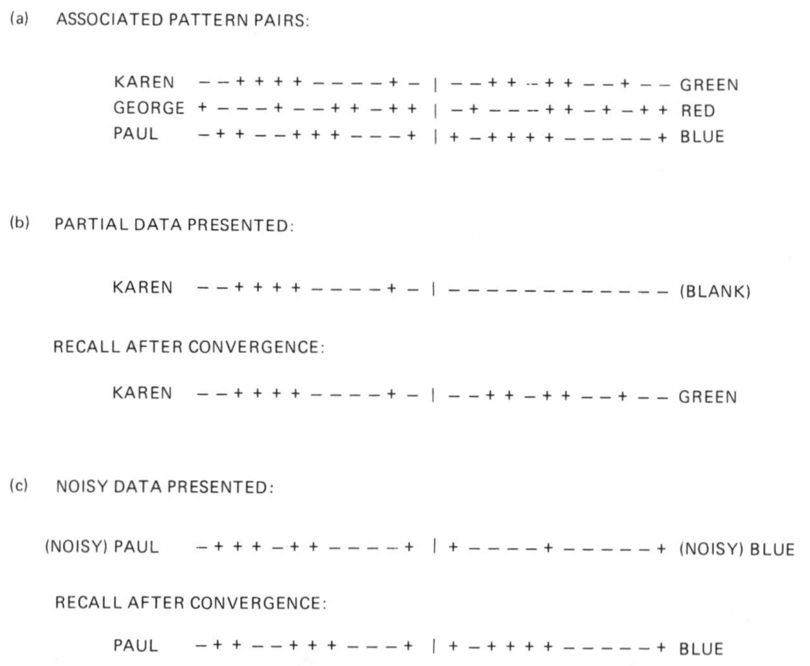

Figure 3-7. Associative memory examples. (a) A 24-unit network stores three associations of names with colors. (b) The name is presented to the network, which recalls the associated color. (c) A noisy pattern pair is presented to the network, which recall the exact pattern pair.

with an initial state that is a noisy version of a memory pattern. The network is expected to converge to the memory that is most accessible to its initial state; if the noisy initial pattern is close enough to the memory pattern, then the network will converge to the correct character. The correct character may then, in principle, be recovered in spite of noise in some of the data.

The pattern in Figure 8b is a *T* with 10% of the pixels incorrect (noisy). When this pattern is presented as an initial state, the network converges to the *T* shown at top left. Figure 8c shows a pattern that started as a *T* but had 20% of its pixels changed. Here the network failed to converge to a *T*, selecting an

Figure 3-8. Pattern completion by a 100-unit Hopfield network that has stored three characters as memories. (a) Three memory patterns. (b) Character T with 10% noise. (c) Character T with 20% noise.

O instead because of the large amount of noise in the initial pattern state. These examples illustrate the pattern completion capacity and limitations of the Hopfield Net.

Setting the Weights

In the associative memory application, the patterns to be stored as memories are chosen a priori. The number of processing units in the network is equal to the number of entries in the patterns to be stored. The weights are fixed based on the choice of patterns.

First we must define notation for the patterns to be stored. Pattern **p**, a vector of +/—'s, such as (++--+-+ ··· +), is denoted by

$$A_p = (a_{p1}, a_{p2}, \ldots, a_{pn})$$

where a_{pi} = *i*th entry in pattern vector **p**. There are m patterns total: A_1, $A_2, \ldots, A_m$.
Then

$$T_{ji} = \sum_{p=1}^{m} \underbrace{(2a_{pi} - 1)}_{1 \text{ or } -1} \underbrace{(2a_{pj} - 1)}_{1 \text{ or } -1} \tag{3-7}$$

where the sum is taken over all patterns to be stored. Note that the term $2a$-1 simply leaves $a = 1$ the same, and changes $a = 0$ to -1. Equation (3-7) accomplishes the following: T_{ji} is incremented by 1 when $a_{pj} = a_{pi}$ (two entries of pattern **p** are the same); T_{ji} is decremented by 1 when $a_{pj} \neq a_{pi}$ (two entries of pattern **p** are different). This increment/decrement process is done for all pairs ij ($i \neq j$) in all patterns $A_{\mathbf{p}}$. One can add a pattern by doing the appropriate increment/decrement process for the new pattern.

Adding pattern memories by this process is analogous to "learning." This is a striking result because Hebb proposed a similar type of learning law with "neural assemblies" in 1949,[8] when it was virtually impossible to justify how such a system might work.

Recall Limitations

Although the Hopfield Net was a significant achievement, there are still many limitations to its actual use. First, the evoked memory is not necessarily the memory pattern that is most similar to the input pattern. A second disadvantage is that all memories are not remembered with equal emphasis; some are evoked inappropriately often. A third disadvantage is that sometimes the network evokes *spurious states*—patterns that were not on the original list of patterns to be stored as memories.

In practical use, the evoked memory may not be the memory that is the closest match to the initial state of the network. (This is true, for example, when the number of matching entries is used as a measure of closeness.) An intuitive description of why is as follows: The initial state of the network defines a point on the "mountainous terrain." This point is on the side of a particular "valley." The minimization procedure finds the bottom of that valley. The bottom point corresponds to the evoked memory. Thus, a memory that is at the bottom of a larger valley will be evoked from a larger number of initial states.

A further practical limitation is the appearance of spurious states. An example of such a state is shown in Figure 3-9. The pairs Walter-White and Walter-Black are both stored in memory in the Hopfield Net. It is possible then to recall Walter-Gray. The reason that Gray is substituted for White is the correlation between the vector for White and the vector for Gray; likewise for the substitution of Black with Gray. Although the network is usually more likely to recall the pairs that were originally encoded (Walter-White and Walter-Black), it is still possible to get the spurious state Walter-Gray. Spurious states have been problematic in other applications of the Hopfield Net as well.

An unlearning process proposed by Hopfield (Hopfield, Feinstein, & Palmer 1983) is intended to address two limitations of the Hopfield Net—the existence of spurious states and the fact that some memories are evoked disproportionately often. In this process, changes in the weights are done according to the equation:

$$\Delta\, T_{ji} = -\epsilon\, (2a_{pj} - 1)\, (2a_{pi} - 1)$$

where ϵ is small, $0 < \epsilon \ll 1.0$.

As a result, this "unlearning" process equalizes the accessibility of memories; it also causes some suppression of spurious states. Although the suppression is not complete, it is significant, as documented by Hopfield and his colleagues (1983).

MEMORY 1:	WALTER	+ + + + – – – –	+ + – + – + – –	WHITE
MEMORY 2:	WALTER	+ + + + – – – –	– – + – + – + +	BLACK
MEMORY 3:	HAROLD	+ + – – + + – –	+ – – + + – – +	GRAY
SPURIOUS STATE:	WALTER	+ + + + – – – –	+ – – + + – – +	GRAY

Figure 3-9. A spurious state in a Hopfield Net associative memory.

There is a maximum limit to the number of memories that a Hopfield Net can store, and this limit increases when the number of units increases. The total number of memories that can be stored is about $0.15N$ memory states, where N is the number of processing units in the network. This limit was determined in experiments with computer simulations (Hopfield 1982) in which memory patterns were generated at random and stored in Hopfield networks. If more such memories are stored, then a significant number of errors occur during recall. Performance can become so poor that memory retrieval is almost impossible. A few errors do occur just below the $0.15N$ limit, and the number of errors goes down if the number of memories is decreased.

The capacity limit of the network depends in part on the choice of memory patterns. If the memory patterns are orthogonal vectors instead of randomly generated vectors, then more memories can be stored. A number of theoretical studies analyze these and other relationships in the capacity of the Hopfield network (Abu-Mostafa & St. Jacques 1985; McEliece et al. 1987).

CONTINUOUS-VALUED HOPFIELD NETWORK

In 1984 Hopfield extended the binary network design to allow for processing units that could attain a continuous range of values. This extension to the binary Hopfield Net is the continuous-valued Hopfield Net. Its activity corresponds directly to the activities in the binary net, but it is capable of doing much more because its units are not limited to binary (0/1) values and its architecture is more complex.

The continuous-valued Hopfield Net retains the same network topology as the binary Hopfield Net. The symmetric weight equation

$$T_{ji} = T_{ij}$$

is also retained. One difference is the use of the sigmoid function instead of the hard threshold given in Eq. (3-2). The sigmoid function provides a continuous-valued, nonlinear—"soft"—threshold.

Figure 3-10 plots the sigmoid function whose equation is

$$f(x) = \frac{1}{(1 + e^{-x})}$$

The soft threshold operates as follows: For values of x sufficiently below a threshold (0), the value of $f(x)$ is almost 0. For values of x sufficiently above a

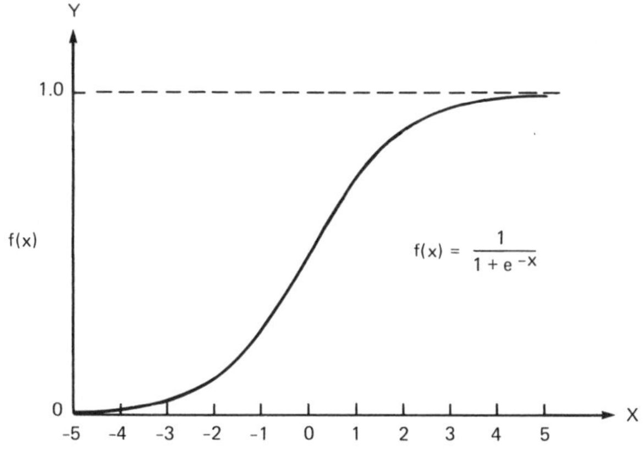

Figure 3-10. The sigmoid function.

threshold (0), the function is close to 1. For a small number of values of x close to 0, the function increases rapidly. The sigmoid can be translated right or left to perform a thresholding at a value other than 0. The sigmoid can be translated or scaled vertically to approach limits other than 0 and 1. (Sigmoid functions also appear in the back-propagation paradigm, see Chapter 4.)

In the continuous-valued Hopfield Net, the architecture of the network is specified so that changes in time are described continuously, rather than as discrete update times for units. The processing units are governed by the equation:

$$C_j \frac{du_j}{dt} = \sum_i T_{ji} V_i - \frac{u_j}{R_j} + I_j \tag{3-8}$$

where C_j = a constant > 0, R_j = controls unit j's decay resistance ($R_j > 0$), I_j = external input to unit j, and V_i is the output of unit i after the sigmoid function is applied.

The energy equation is

$$E = -\frac{1}{2} \sum_j \sum_i T_{ji} u_j u_i - \sum_j u_j I_j \tag{3-9}$$

where I_j is an input bias term. For applications, T_{ji} and I_j are chosen appropriately so that E represents the function that is minimized to solve the applications problem.

An Optimization Example: The Traveling Salesman Problem

The traveling salesman problem (TSP) is an extremely difficult classic optimization problem. This problem is in the class of *NP*-complete problems, all of which take a very long time to solve in worst-case analysis. Hopfield and Tank devised a way to attack this problem using the continuous Hopfield network, which found a good solution to the traveling salesman problem in a reasonable amount of time.

Unfortunately, practical limitations exist: Performance is not good with the Hopfield Net, and performance becomes poorer with larger problems (Wilson & Pawley 1988). Nevertheless, the behavior of the network is interesting to observe.

The TSP is stated as follows: A salesman has a number of cities to visit. He starts and ends at a particular city, and travels to each city on his list without visiting any city twice. The goal is to choose the order of the cities visited in such a way that the total path length traveled is minimized.

Let the cities to be visited be $A, B, C, \ldots, T$, with the distances between them given as

$$d_{xy} = \text{distance from city } x \text{ to city } y$$

A "tour" is a trip that starts and ends at the same place and visits each city exactly once. The path length of the tour $X_1 X_2 \ldots X_t$ is

$$\sum_{t=1}^{n} d_{X_t X_{t+1}} \qquad (3\text{-}10)$$

where the subscripts of X are evaluated modulo n.

The minimum solution to the traveling salesman problem is the tour that has the smallest value for the total distance traveled. A "good" solution is a tour that is relatively close to the minimum solution. Hopfield was able to get a good solution rather than the minimum solution.

The first key to addressing this problem with the Hopfield network is the representation of the tour. A tour can be represented by a matrix of 0s and 1s. The rows of the matrix represent the different cities, and the columns of the matrix are the different positions in a tour (e.g., the first city is in position 1, etc.). A matrix representation for a tour of 10 cities is given in Figure 3-11.

A Hopfield Net can be built with the same number of processing units as there are entries in this matrix. Thus, a 10-city TSP would require a Hopfield Net with 100 (10^2) units. The goal of this application is to get the values of the 100 units in the Hopfield Net to converge to become the values for a matrix that represents a tour that is a good solution of the TSP.

POSITION IN TOUR

CITY	1	2	3	4	5	6	7	8	9	10
A	0	0	0	1	0	0	0	0	0	0
B	0	1	0	0	0	0	0	0	0	0
C	0	0	0	0	0	0	1	0	0	0
D	0	0	0	0	1	0	0	0	0	0
E	0	0	0	0	0	0	0	0	0	1
F	0	0	0	0	0	0	0	1	0	0
G	1	0	0	0	0	0	0	0	0	0
H	0	0	0	0	0	1	0	0	0	0
I	0	0	0	0	0	0	0	0	1	0
J	0	0	1	0	0	0	0	0	0	0

TOUR: G B J A D H C F I E

Figure 3-11. The traveling salesman matrix, specifying the tour at bottom.

The second key to solving the TSP with the Hopfield network is in the energy equation. The energy equation used is complex, and can be rearranged to match the terms in the original energy equation (3-9) given by Hopfield.

The energy equation for the TSP is:

$$E = (A/2) \sum_{X} \sum_{i} \sum_{\substack{j \\ j \neq i}} V_{Xi} V_{Xj} \qquad \text{(term 1)}$$

$$+ (B/2) \sum_{i} \sum_{X} \sum_{\substack{Y \\ Y \neq X}} V_{Xi} V_{Yi} \qquad \text{(term 2)}$$

$$+ (C/2) (\sum_{X} \sum_{i} V_{Xi} - n^2) \qquad \text{(term 3)}$$

$$+ (D/2) \sum_{X} \sum_{Y} \sum_{i} d_{XY}\, u_{Xi} (V_{X,i+1} + V_{Y,i-1}) \qquad \text{(term 4)}$$

where A, B, C, and D are parameter values that may be tuned to aid performance and $V_{X,i}$ is the entry in the table, row X, column i.

Each term in the energy function may be explained intuitively as imposing a particular restriction on the tour of the salesman, and each restriction is enforced through the minimization of the corresponding term. Ideally, if the constants are sufficiently large, then all of the low-energy states of the network will have the form of a valid tour, and the states with the shortest paths will be the lowest-energy states.

The first term allows only one visit to each city; this term is small when there is only a single 1 in a given row. The second term does not allow the salesman to be in two different cities at the same time; this term is small when there is only a single 1 in a column. The third term allows only n cities to appear on the itinerary; this term is small when there are only n 1s in the matrix. The fourth term expresses the distance sum to be minimized; this last term is a mathematical expression proportional to the total distance of the tour.

The weights are set initially by Eq. (3-11), which takes into account the distances between the cities

$T_{Xi,Yj} =$

$-A\,\delta_{XY}\,(1-\delta_{ij})$	inhibiting connections within each row
$-B\,\delta_{ij}\,(1-\delta_{XY})$	inhibitory connections within each column
$-C$	global inhibition
$-D\,d_{XY}\,(\delta_{j,i+1}+\delta_{j,i-1})$	distance term

The external bias is

$$I_{Xj} = Cn$$

We use the notation:

$$\delta_{ij} = \begin{cases} 1 & \text{if } i = j \\ 0 & \text{otherwise} \end{cases}$$

The weights remain fixed at the values given in Eq. (3-10) throughout the optimization procedure. The initial state of the network is set at random, and updating is performed until the network converges to a steady state. If convergence has been successful, the itinerary of the salesman is then read from the final state of the network.

In Figure 3-12, which showed the convergence of the network for one example run, the value of the matrix entry is proportional to the diameter of each circle. Thus, larger circles correspond to larger entries in the matrix of Figure 3-11. Figure 3-12a–c shows snapshots of intermediate times; Figure 3-12d shows the final state of the network, after it converged to a stable state.

Figure 3-12e also depicts an intermediate state of the network, the route of the salesman's travel plan. Intermediate states of the network can sometimes show the itinerary going to two places at the same time (two entries in the same column), or may omit visiting certain cities at all (no entries in a given row). The intermediate states may appear to be "considering" different possibilities.

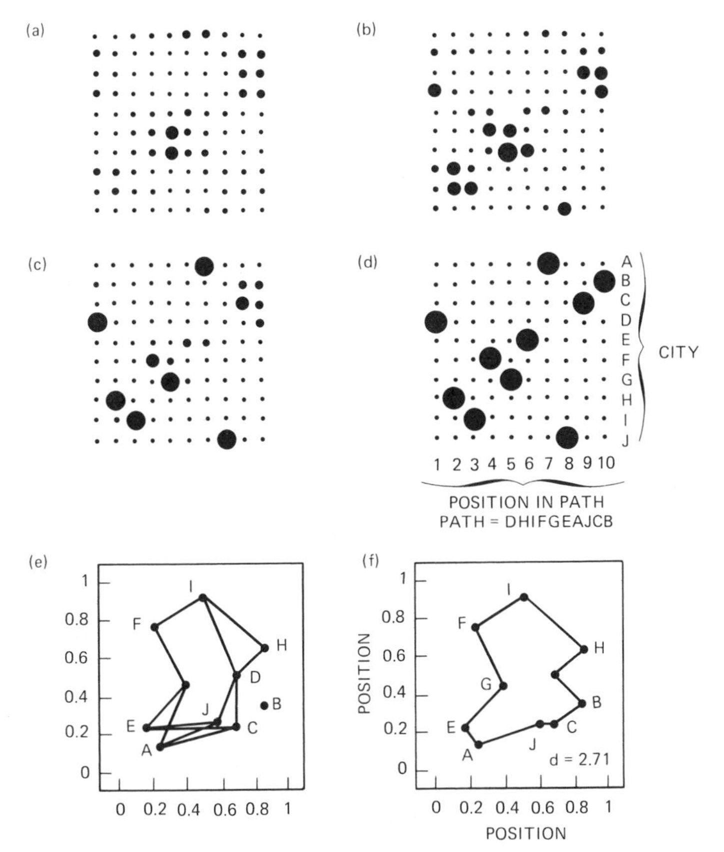

Figure 3-12. A traveling salesman solution. (a,b,c) Progressive intermediate states of the network. (d) Final state of the network. (e) Incomplete tours indicated by an intermediate state. (f) Final tour, indicated from the final state of the network shown in (c) (from Hopfield and Tank. Neural computation of decisions in optimization problems, *Biological Cybernetiks*. Springer-Verlag 1985).

The tour that corresponds to the final state of the network, Figure 3-12f, obeys the basic rules of visiting each city, with each city visited only once. Furthermore, the distance is relatively short compared to other possible tours.

Although Hopfield and Tank reported success with the traveling salesman problem, their work was apparently limited to a subset of problems involving fewer than 30 cities. Other investigators (Wilson & Pawley 1988) have reported that the Hopfield network does not solve larger problems (with more than 10 cities) reliably. However, in the larger problems the network often comes up with a solution that contains segments that are locally optimal.

ADVANTAGES AND LIMITATIONS

In each of these uses of the Hopfield network for associative memory and optimization, the applications problem is solved when the network reaches a stable state at an energy minimum. Such an approach was striking when proposed because of its use of an interconnected net of processing units that "settle" or converge on a final solution.

Applications of the Hopfield Net are unfortunately limited by the network's level of performance: Associative memory has displayed limited capacity, uneven recall ability, and recall of extra (spurious) states. Optimization of the traveling salesman problem provides a limited number of good answers and does not work well for more than 10 cities. The Hopfield network nevertheless gives an excellent demonstration of problems that can be attacked with neural networks. Realistic real-world applications, however, will probably require enhanced network architectures.

Hopfield and Tank have shown additional examples of the Hopfield Net in a variety of optimization problems, including signal decomposition and linear programming. Some job shop scheduling optimization has been addressed with a Hopfield-like network (Foo and Takefuji 1988). Such optimization problems, however, tend to work on examples from a limited domain — the problems solved are not general.

Many possible applications exist for the Hopfield network and similar, more sophisticated architectures, such as the Boltzmann Machine. Candidates include speech processing, database retrieval, image processing, fault tolerant memories, and pattern classification.

The nonsynchronous updating of processing units in the Hopfield network is a unique property, and is especially relevant to the study of biological systems because of their asynchronous updating of nerve cell attributes. In addition, asynchronous updating can be helpful in designing fast hardware implementations.

The most general advantage of the Hopfield network is its inherently parallel architecture. As a result, potential hardware implementations may be very fast. Tradeoffs must be assessed, however, between the size and speed of the network and the size of the applications problems.

References

Abu-Mostafa, Y. S. and J. St. Jacques. 1985. *IEEE Trans. Info. Theory* 31(4): 461–64.

Anderson, James A. 1972 "A Simple Neural Network Generating an Interactive Memory," *Mathematical Biosciences*, Volume 14, pp 197–220, 1972.

Anderson, James A., Silverstein, Jack W., Ritz, Stephen A., Jones, Randall S. 1977 "Distinctive Features, Categorical Perception, and Probability Learning: Some

Applications of a Neural Model," *Psychological Review*, Volume 84, Number 5, September 1977.

Foo, Y. S. and Y. Takefuji. 1988. Stochastic neural networks for solving job-shop scheduling: Parts 1 and 2. *IEEE ICNN Proc. 1988* II-275–II-290.

Fukushima, K., S. Miyake, and T. Ito. 1983. Neocognitron: A neural network model for a mechanism of visual pattern recognition. *IEEE Trans. Systems, Man, and Cybernetics* 13(5): 826–34.

Grossberg, S. 1987. *The Adaptive Brain,* Vols. 1 & 2. New York: North-Holland.

Grossberg, S. 1988. *Neural Networks and Natural Intelligence.* Cambridge, Mass.: MIT Press.

Hebb, D. O. 1949. *Organization of Behavior,* New York: Wiley.

Hinton, G. E. and T. J. Sejnowski. 1986. Learning and relearning in Boltzmann Machines. In *Parallel Distributed Processing.* Cambridge, Mass.: 282–317. MIT Press.

Hopfield, J. J. 1982. Neural networks and physical systems with emergent collective computational abilities. *Proc. Natl. Acad. Sci.* 79: 2554–58.

Hopfield, J. J. 1984. Neurons with graded response have collective computational properties like those of two-state neurons. *Proc. Natl. Acad. Sci.* 81: 3088–3092.

Hopfield, J. J., D. I. Feinstein, & R. G. Palmer. 1988. Unlearning has a stabilizing effect in collective memories. *Nature* 304: 158–59.

Hopfield, J. J. and D. W. Tank. 1985. Neural computation of decisions in optimization problems. Biol. Cybernetics. 52: 141–152.

McEliece, R. J., E. C. Posner, E. R. Rodemich, and S. S. Venkatesh. July 1987. The capacity of the Hopfield associative memory. *IEEE Trans. Info. Theory* 33(4): 461–482.

Reilly, D. L., L. N. Cooper, and C. Elbaum, 1982. "A Neural Model for Category Learning." Biological Cybernetics. 45, pp. 35–41.

Wilson, G. V. and G. S. Pawley. 1988. On the stability of the traveling salesman problem algorithm of Hopfield and Tank. *Biol. Cybern.* 58: 63–70.

Suggested Readings

Anderson, James A. 1983 "Cognitive and Psychological Computation with Neural Models", IEEE Transactions on Systems, Man, and Cybernetics, Volume SMC-13, Number 5, September/October 1983.

Hopfield, J. J. and D. W. Tank. 1986. Computing with neural circuits: A model. *Science* 233:625–33.

Tank, D. W. and J. J. Hopfield. 1986. Simple neural optimization networks. *IEEE Trans. CS.* CAS-33(5): 533–41.

Van den Bout, D. E. and T. K. Miller. 1988. A traveling salesman objective function that works. *IEEE ICNN 1988* II-299–II-304.

4

Back-Error Propagation

Back-error propagation is the most widely used of the neural network paradigms and has been applied successfully in applications studies in a broad range of areas. Applications studies have spanned tasks from military pattern recognition to medical diagnosis, and from speech recognition and synthesis to robot and autonomous vehicle control. Back-propagation has been applied to character recognition, sonar target recognition, image classification, signal encoding, knowledge processing, and a variety of other pattern-analysis problems. Back-propagation can attack any problem that requires pattern mapping: Given an input pattern, the network produces an associated output pattern.

Back-propagation is one of the easiest networks to understand. Its learning and update procedure is intuitively appealing because it is based on a relatively simple concept: If the network gives the wrong answer, then the weights are corrected so that the error is lessened and as a result future responses of the network are more likely to be correct.

Back-propagation networks are usually layered, with each layer fully connected to the layers below and above. When the network is given an input, the updating of activation values propagates forward from the input layer of processing units, through each internal layer, to the output layer of processing units. The output units then provide the network's response. When the network corrects its internal parameters, the correction mechanism starts with the output units and back-propagates backward through each internal layer to the input layer—hence the term back-error propagation, or back-propagation.

The conceptual basis of back-propagation was first presented in 1974 by Paul Werbos, then independently reinvented by David Parker in 1982, and presented to a wide readership in 1986 by Rumelhart and McClelland. Their book, *Parallel Distributed Processing* introduced the broad potential of the neural network approach and widespread interest in back-propagation followed.

Notable early applications of back-propagation were done by Terry Sejnowski and his colleagues at Johns Hopkins University. Particularly striking is the NETTalk program, by Sejnowski and Rosenberg, in which a back-propagating neural network was trained in the rules of phonetics in just two weeks of CPU time on a VAX. The network, which produced sounds as it was learning phonetic rules, sounded uncannily like a child learning to read aloud.

Back-propagation is a tremendous step forward compared to its predecessor, the perceptron. The perceptron was limited to only two layers of processing units, with only a single layer of adaptable weights. This key limitation meant that the perceptron could only classify patterns that were linearly separable. Back-propagation overcomes this limitation because it can adapt two or more layers of weights, and uses a more sophisticated learning rule. The power of back-propagation lies in its ability to train hidden layers and thereby escape the restricted capabilities of single-layer networks.

When two or more layers of weights are adjusted, the network has middle — or hidden — layers of processing units. Each hidden layer acts as a layer of "feature detectors" — units that respond to specific features in the input pattern. These feature detectors organize as learning takes place, and are developed in such a way that they accomplish the specific learning task presented to the network.

For years it has been known that the key to pattern-recognition problems and many learning tasks has been in the choice of salient features to consider. Knowing the correct distinguishing features allows one to classify different patterns. Back-error propagation can identify an effective set of features automatically. Thus, a fundamental step toward solving pattern recognition problems has been taken with back-error propagation.

AN OVERVIEW

Typically, back-propagation employs three or more layers of processing units. Figure 4-1 shows the topology for a typical three-layer back-propagation network. The bottom layer of units is the input layer — the only units in the network that receive external input. The layer above is the hidden layer, in which the processing units are interconnected to layers above and below. The top layer is the output layer.

The layers in Figure 4-1 are fully interconnected — each processing unit is connected to every unit in the layer above and in the layer below. Units are not

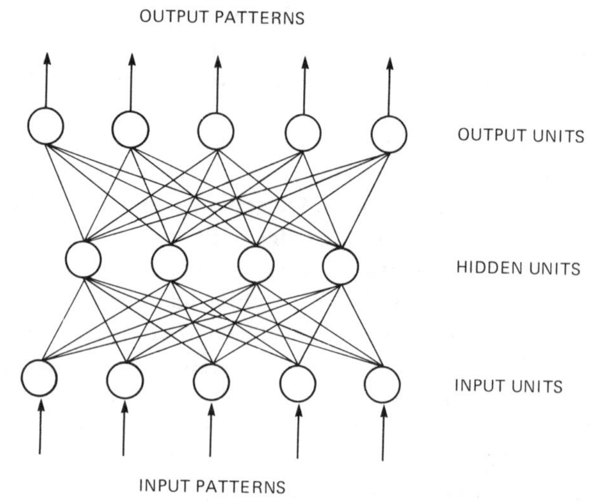

Figure 4-1. A three-layered back-propagation network, fully interconnected.

connected to other units in the same layer. A back-propagation network must have at least two layers. Figure 4-2 shows such a network with five layers, all fully interconnected. The three internal layers are hidden layers (with hidden units). Back-propagation networks do not have to be fully interconnected,

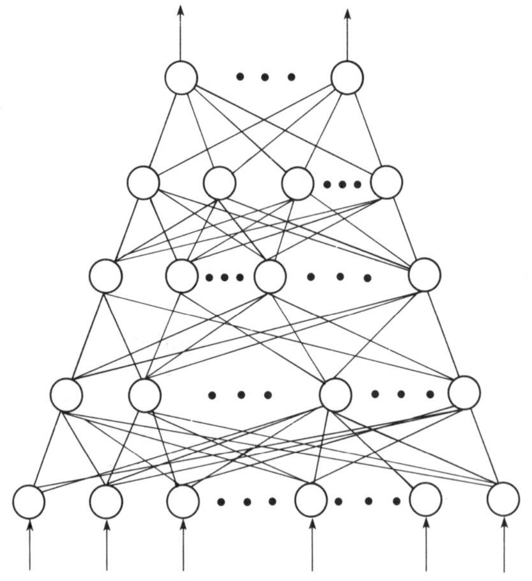

Figure 4-2. A five-layered back-propagation network, fully interconnected.

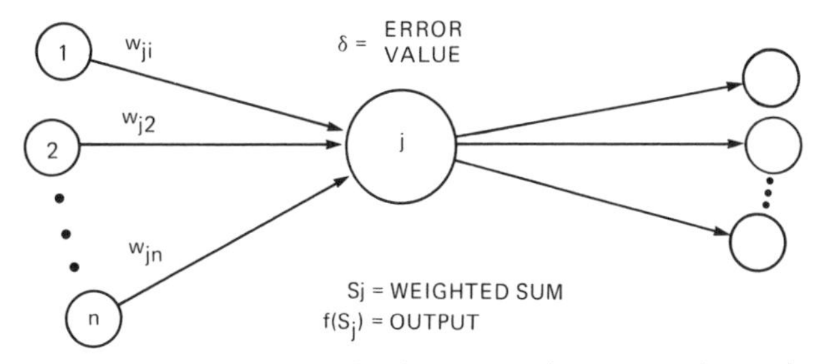

Figure 4-3. The basic back-propagation processing unit.

although most applications work has been done with fully interconnected layers.

In Figure 4-3, a basic back-propagation processing unit, inputs are shown at the left, and at the right are units that receive outputs from the processing unit at the center. The processing unit has a weighted sum of inputs (S_j), an output value (a_j), and an associated error value (δ_j) that is used during weight adjustments.

Weights associated with each interconnection are adjusted during learning. The weight to unit j from unit i is denoted here as w_{ji}. After learning is completed, the weights are fixed. These final values are then used during "recall" sessions. Figure 4-3 illustrates the weights along the incoming connections to the processing unit at the center. There is a matrix of weight values

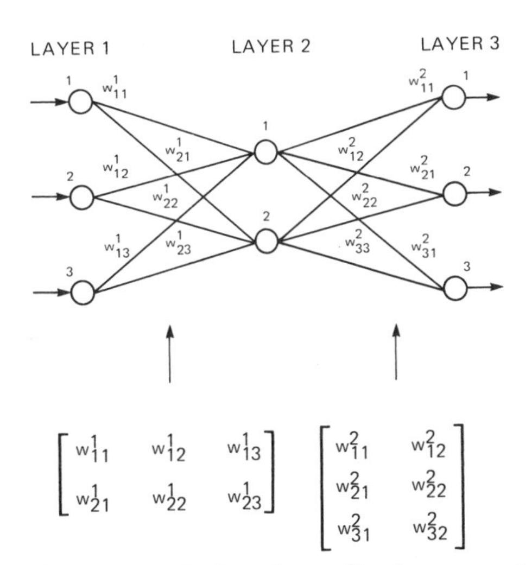

Figure 4-4. Weight matrices of a three-layered back-propagation system. Units are indexed starting with 1 in each layer. Superscripts have been added to distinguish weights in different layers.

that corresponds to each layer of interconnections (Figure 4-4); these matrices are indexed with superscripts to distinguish weights in different layers.

The *activation level* parameter associated with each processing unit in the back-propagation network is also the output value of each processing unit. An animated simulation of a back-propagation neural network depicts the activation level for each unit by using colors, gray levels, or different sizes of units.

Figure 4-5 graphically shows both the activation levels and weight levels in a back-propagation network. Squares of different shades represent processing units with different activation levels. The two grids represent two layers of weights, with the shade of each square determined by the weight value for each interconnection. The numeric scale for determining the shade is different for the activation levels and the weights, as each type of parameter has a different range.

A back-propagating neural network is trained by supervised learning. The network is presented with pairs of patterns—an input pattern paired with a target output. Upon each presentation, weights are adjusted to decrease the difference between the network's output and the target output. A training set—a set of input/target pattern pairs—is used for training, and is presented to the network many times. After training is stopped, the performance of the network is tested.

The back-propagation learning algorithm involves a forward-propagating step followed by a backward-propagating step. Both the forward- and back-propagation steps are done for each pattern presentation during training. The forward-propagation step begins with the presentation of an input pattern to the input layer of the network, and continues as activation level calculations propagate forward through the hidden layers. In each successive layer, every processing unit sums its inputs and then applies a sigmoid function to compute its output. The output layer of units then produces the output of the network.

The backward-propagation step begins with the comparison of the network's output pattern to the target vector, when the difference, or "error," is calculated. The backward-propagation step then calculates error values for hidden units and changes for their incoming weights, starting with the output layer and moving backward through the successive hidden layers. In this back-propagating step the network corrects its weights in such a way as to decrease the observed error.

The error value (δ) associated with each processing unit reflects the amount of error associated with that unit. This parameter is used during the weight-correction procedure, while learning is taking place. A larger value for δ indicates that a larger correction should be made to the incoming weights, and its sign reflects the direction in which the weights should be changed.

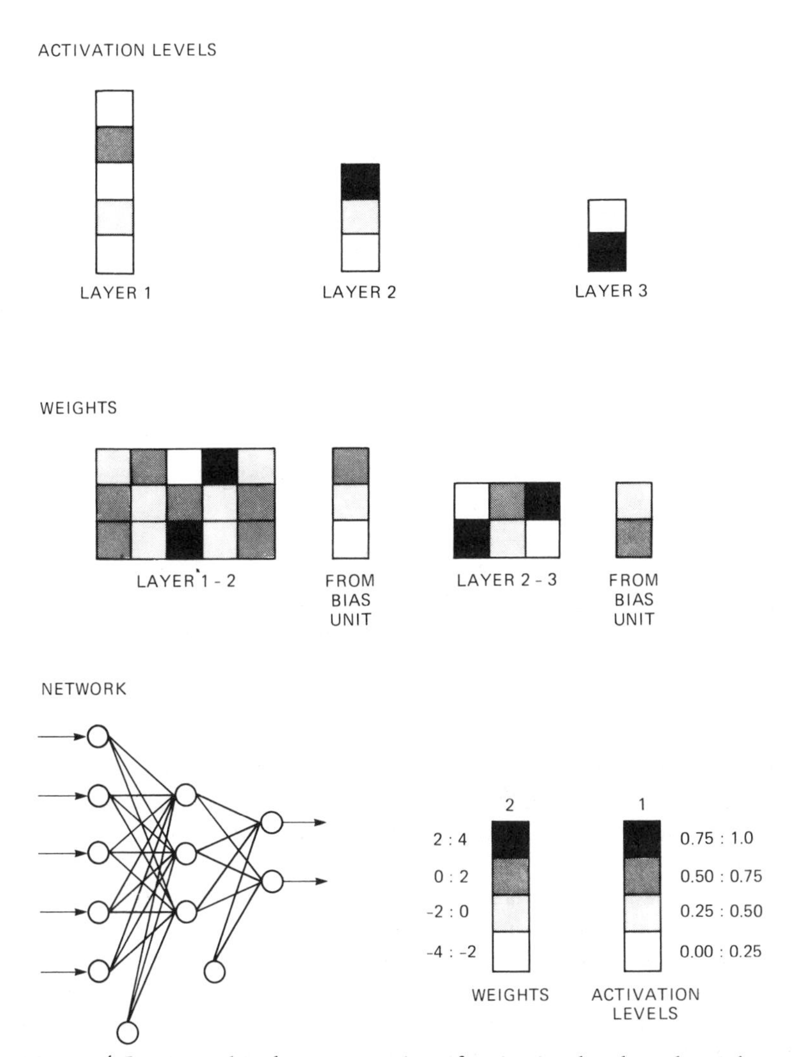

Figure 4-5. A graphical representation of activation levels and weights.

FORWARD-PROPAGATION

The forward-propagation step is initiated when an input pattern is presented to the network. Each input unit corresponds to an entry in the input pattern vector, and each unit takes on the value of this entry. After the activation levels for the first layer of units is set, the remaining layers perform a forward-propagation step, which determines the activation levels of the other layers of units.

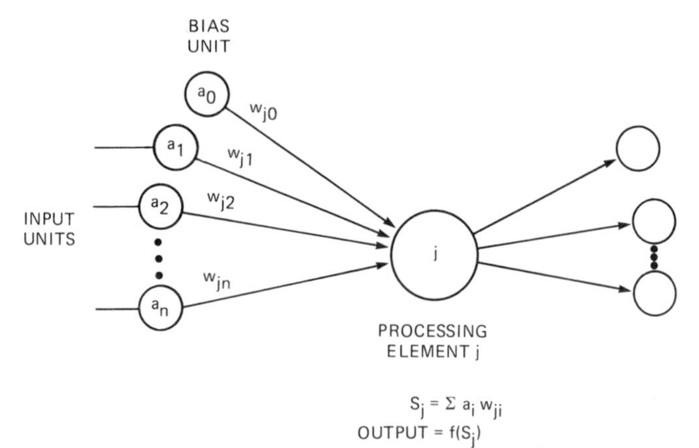

Figure 4-6. The forward-propagation step.

Figure 4-6 illustrates the specifics of the forward-propagation step. Incoming connections to unit j are at the left and originate at units in the layer below. Output values from these units arrive at unit j and are summed by

$$S_j = \sum_i a_i w_{ji} \tag{4-1}$$

where a_i = the activation level of unit i, and w_{ji} = the weight from unit i to unit j (unit i is one layer below unit j). After the incoming sum S_j is computed, a function f is used to compute $f(S_j)$. The function f, a sigmoid curve, is illustrated in Figure 4-7a.

The sigmoid curve is relatively flat at both ends, and has a rapid rise in the middle. When x is less than -3, $f(x)$ is close to 0; when x is greater than 3, $f(x)$ is close to 1. In fact, $f(x)$ approaches 1 asymptotically as x gets larger, and $f(x)$ approaches 0 asymptotically as x becomes a greater negative value. There is a transition from 0 to 1 that takes place when x is approximately 0 ($-3 < x < 3$). The sigmoid function performs a sort of "soft" threshold that is rounded (and differentiable) compared to the step function (Figure 4-7b).

The equation for the sigmoid function is

$$f(x) = \frac{1}{1 + e^{-x}} \tag{4-2}$$

Since the operand is the weighted sum of unit j, we have

$$f(S_j) = \frac{1}{1 + e^{-S_j}} = \frac{1}{1 + e^{-\Sigma a_i w_{ji}}}$$

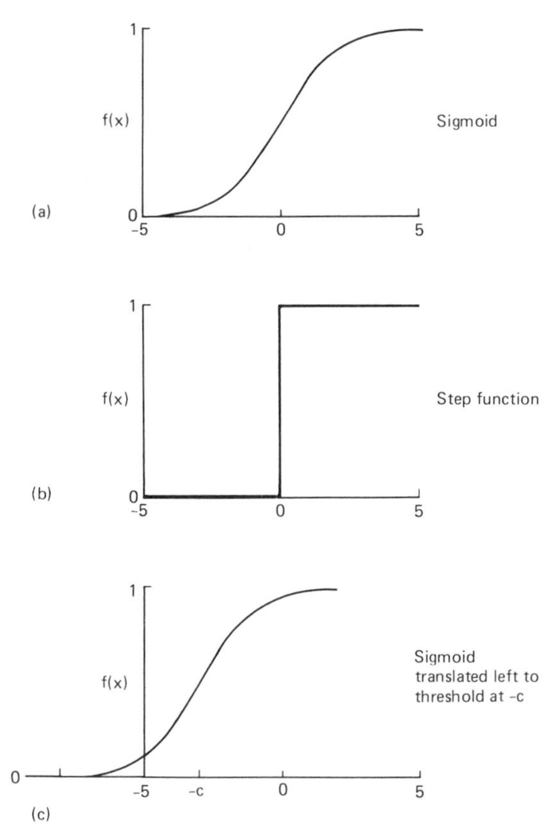

Figure 4-7. Threshold functions. (a) The sigmoid function. (b) A step function. (c) A sigmoid moved *c* units to the left, to threshold at $-c$.

After the sigmoid function is computed on S_j, the resulting value becomes the activation level of unit *j*. This value, the output of unit *j*, is sent along the output interconnections (on the right of Figure 4-6). The same output value is sent along all of the output interconnections.

The input layer of units is a special case. These units do not perform the weighted sum on their inputs because each input unit simply assumes the corresponding value taken from the input vector. We consider the input layer to be a layer of the network even though it does not perform the weighted sum and sigmoid calculations.

Some back-propagation networks employ a bias unit as part of every layer but the output layer. This unit has a constant activation value of 1. Each bias unit is connected to all units in the next higher layer, and its weights to them are adjusted during the back-error propagation. The bias units provide a constant term in the weighted sum of the units in the next layer. The result is sometimes an improvement on the convergence properties of the network.

The bias unit also provides a "threshold" effect on each unit it targets. It contributes a constant term in the summation S_j, which is the operand in the sigmoid function (4-2). This is equivalent to translating the sigmoid curve in Figure 4-7 to the left or to the right. For example, suppose the bias unit (a_0) in Figure 4-6 has an output value of 1.0 and a weight

$$C = w_{j0}$$

Then let

$$z = \sum_{i=1}^{n} a_i w_{ji}$$

Then z is the incoming sum from all of the units in Figure 4-6 other than the bias unit. If the bias unit contributes the constant C to the incoming sum of unit j, then this sum becomes

$$z + C$$

Now compare the graph of the sigmoid function for $f(z)$, given in Figure 4-7a, to the graph of the function $f(z + C)$, shown in Figure 4-7c. The constant C translates the graph to the left by the amount C, thus moving the threshold of the sigmoid curve from 0 to $-C$. In this way, the bias units provide an adjustable threshold for each target unit. The threshold for unit j then comes from the value of w_{j0}, the weight of the interconnection from the bias unit.

BACKWARD PROPAGATION

Figure 4-8 illustrates the backward propagation step. Here the δ values are calculated for all processing units and weight changes are calculated for all interconnections. The calculations begin at the output layer and progress backward through the network to the input layer.

The error-correction step takes place after a pattern is presented at the input layer and the forward-propagation step is complete. Each processing unit in the output layer produces a single real number for its output, which is compared to the target output specified in the training set (Figure 4-8a). Based on this difference, an error value is calculated for each unit in the output layer as in Fig. 4-8b. Then the weights are adjusted for all of the interconnections that

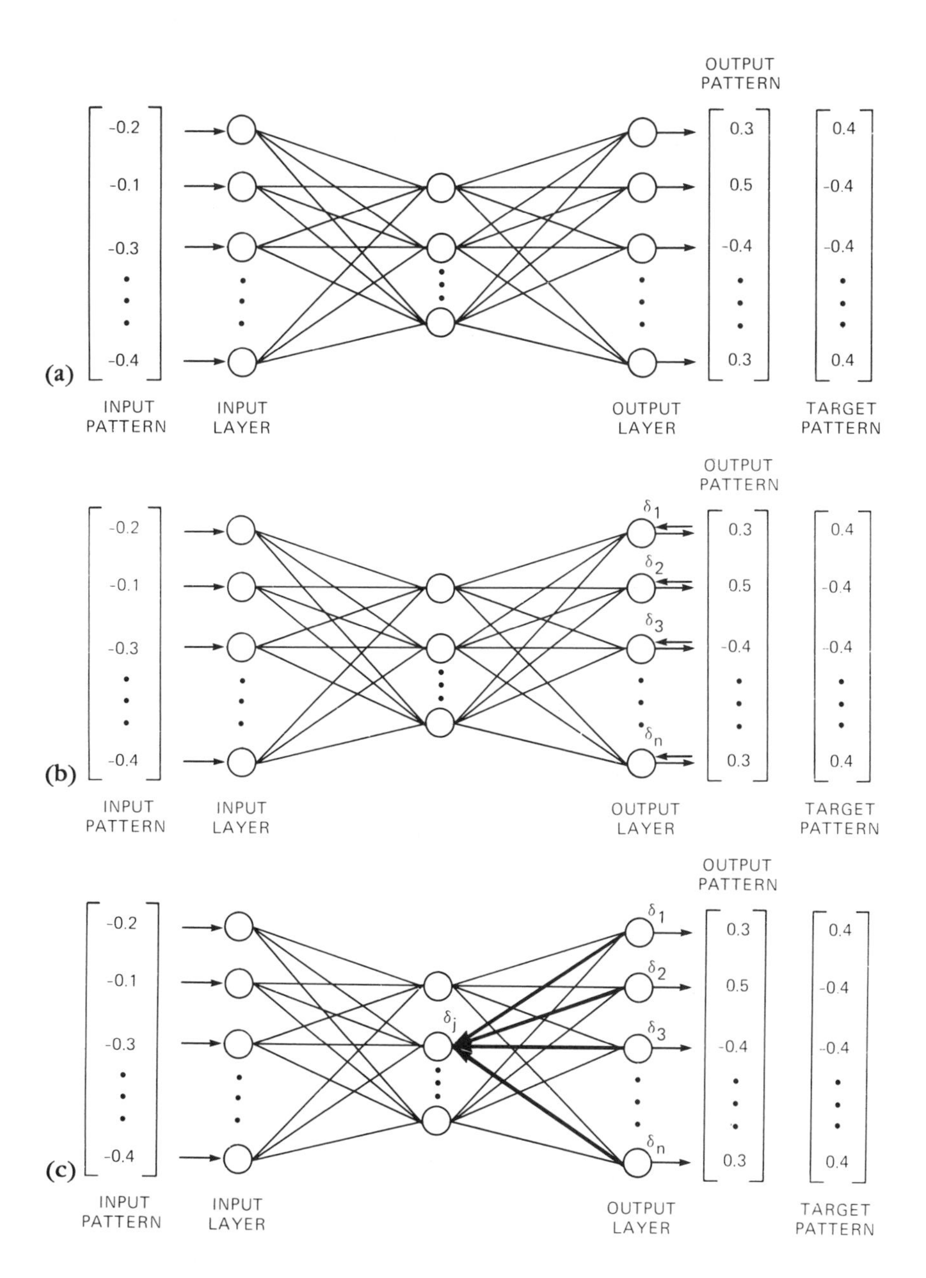

Figure 4-8. Basic back-propagation dynamics. (a) After forward propagation, the target pattern is compared to the output pattern. (b) δ values are calculated for the output layer. Arrows represent flow of information. After δ values are calculated for the output layer, its incoming weights are adjusted. (c) δ values are calculated for the hidden layer. Heavy lines indicate that δ values are communicated from the output layer to the hidden layer. After δ values are calculated for the hidden layer, its incoming weights are adjusted.

go into the output layer. Next an error value is calculated for all of the units in the hidden layer that is just below the output layer (Fig. 4-8c). Then the weights are adjusted for all interconnections that go into the hidden layer. The process is continued until the last layer of weights has been adjusted.

The error value, denoted by the variable δ, is simple to compute for the output layer and somewhat more complicated for the hidden layers. If unit j is in the output layer, then its error value is:

$$\delta_j = (t_j - a_j)f'(S_j) \tag{4-3}$$

where

t_j = the target value for unit j
a_j = the output value for unit j
$f'(x)$ = the derivative of the sigmoid function f
S_j = weighted sum of inputs to j

The quantity $(t_j - a_j)$ reflects the amount of error. The f' part of the term "scales" the error to force a stronger correction when the sum S_j is near the

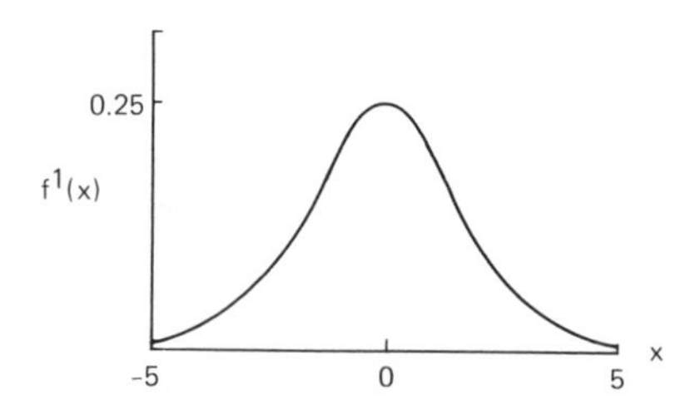

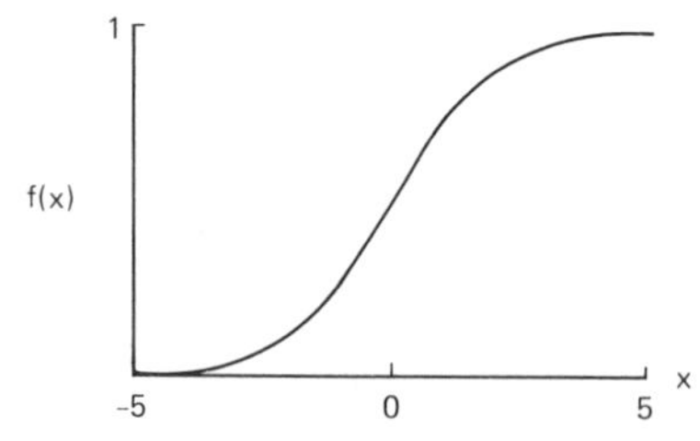

Figure 4-9. The sigmoid function (below) compared to its derivative (above).

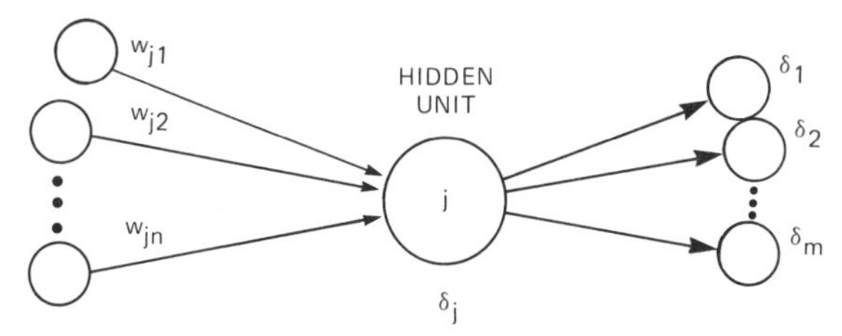

Figure 4-10. A processing unit in a hidden layer.

rapid rise in the sigmoid curve. Figure 4-9 illustrates the form of the f' function, with a peak in the same position as the rise in the sigmoid curve.

Figure 4-10 illustrates j as a unit in a hidden layer. In such a situation, the error value of j is computed as:

$$\delta_j = \left[\sum_k \delta_k w_{kj} \right] f'(S_j)$$

In this case, a weighted sum is taken of the δ values of all units that receive output from unit j. The f' again serves to "scale" this output by emphasizing the region of rapid rise of the sigmoid function.

The adjustment of the connection weights is done using the δ values of the processing unit. Each interconnection weight is adjusted by taking into account the δ value of the unit that receives input from that interconnection. The connection weight adjustment is done as follows:

$$\Delta w_{ji} = \eta \delta_j a_i \tag{4-4}$$

Figure 4-11 diagrams the adjustment of weight w_{ji}, which goes to unit j from unit i. The amount adjusted depends on three factors: δ_j, a_i, and η. This weight adjustment equation is known as the generalized δ rule (Rumelhart & McClelland 1986).

The size of the weight adjustment is proportional to δ_j, the error value of the target unit. Thus a larger error value for unit j results in larger adjustments to its incoming weights.

The weight adjustment is also proportional to a_i, the output value for the originating unit. If this output value is small, then the weight adjustment is small. If this output value is large, then the weight adjustment is large. Thus a

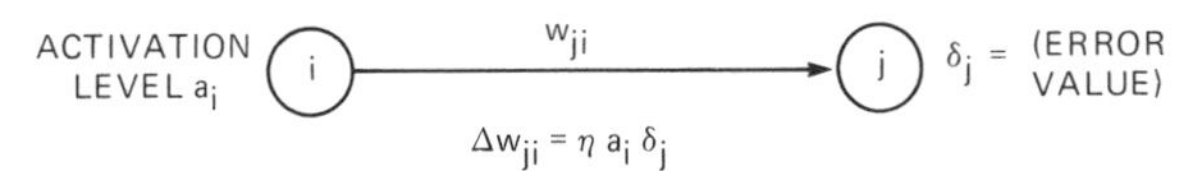

Figure 4-11. Updating a weight.

higher activation value for incoming unit *i* results in a larger adjustment to its outgoing weight.

The variable η in the weight-adjustment equation is the ***learning rate.*** Its value—commonly between 0.25 and 0.75—is chosen by the neural network user, and usually reflects the rate of learning of the network. Values that are very large can lead to instability in the network, and unsatisfactory learning. Values that are too small can lead to excessively slow learning. Sometimes the learning rate is varied in an attempt to produce more efficient learning of the network; for example, allowing the value of η to begin at a high value and to decrease during the learning session can sometimes produce better learning performance.

NETWORK TRAINING

Back-propagation networks are trained by a technique called ***supervised learning,*** whereby the network is presented with a series of pattern pairs—each pair consisting of an input pattern and a target output pattern. Each pattern is a vector of real numbers. The target output pattern is the desired response to the input pattern and is used to determine the error values in the network when the weights are adjusted.

The target output pattern is sometimes designed to represent a classification for the input pattern. In this way, the network may be presented with a series of input patterns together with the classification for each input pattern. In other applications, the target output is simply a desired pattern response to the input pattern, and the network is trained to be a pattern-mapping system.

Figure 4-12 shows an example training set that may be used for a back-propagation network. The input patterns are shown along with their corresponding target output patterns. Each pattern is a vector of three numbers. The neural network to be trained, shown at the bottom, has three processing units in the input layer and three processing units in the output layer.

The training task in this case is to learn to classify the three different types of graphs shown at the left. The target outputs require that a different output unit become active with each different type of graph. Thus, the target outputs reflect the correct classification for each input pattern.

The patterns in the training set are presented to the network repeatedly. Each training iteration consists of presenting each input/output pattern pair once. When all patterns in the training set have been presented, the training iteration is completed, and the next training iteration is begun. A typical back-propagation example might entail hundreds or thousands of training iterations.

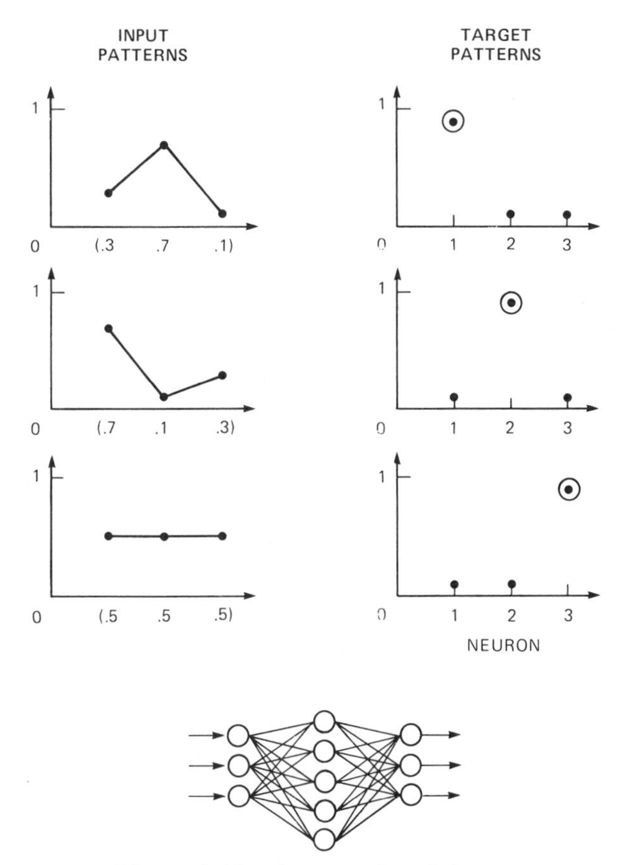

Figure 4-12. An example training set.

CONVERGENCE

When a network is trained successfully, it produces correct answers more and more often as the training session progresses. It is important, then, to have a quantitative measure of learning. The root-mean-squared (RMS) error is usually calculated to reflect the degree to which learning has taken place in the network [Eq. (2-5)]. This measure reflects how close the network is to getting the correct answers. As the network learns, its RMS error decreases (as in Figure 2-5). Generally, an RMS value below 0.1 indicates that a network has learned its training set.

Note that whether an answer is correct or not is a binary yes/no decision.

The target value to a network is a real number, and so is the output value. Thus the network does not provide a yes/no response that is either "correct" or "incorrect." The network gets closer and closer to the target value incrementally with each step. It is possible then to define a cutoff point when the network's output is said to match the target values, and allow this to define a "correct" answer.

Convergence is a process whereby the RMS value for the network gets closer and closer to 0. Convergence is not always easy to achieve because the process may take an exceedingly long time and sometimes the network gets stuck in a local minimum and stops learning altogether.

It is possible to represent convergence intuitively in terms of walking about on mountainous terrain. The terrain is the graph of the RMS value as a function of all of the weights in the network. Using this analogy, the back-propagation algorithm is seeking a minimum height in this mountainous terrain.

Ideally, we seek a global minimum—the bottom of the valley that is the lowest in the entire terrain. This corresponds to the lowest RMS value possible. Unfortunately, it is possible to encounter a local minimum—a valley that is not the lowest possible in the entire terrain. Nevertheless, a local minimum is surrounded by higher ground, and the network usually does not leave a local minimum by the standard back-propagation algorithm described. Special techniques should be used to get out of a local minimum.

The appearance of a local minimum is not always a significant problem. Back-propagation networks typically converge to a good RMS value when the training examples are clearly distinguishable. When a local minimum is encountered, the network may be able to avoid entering that local minimum by a number of techniques, for example, changing the learning parameter or the number of hidden units. These techniques tend to change the scenario involved with moving about on the "mountainous terrain" and may cause the network to avoid the local minimum.

Adding small random values to the weights allows the network to escape from a local minimum once it is encountered by moving the position of the network from a local minimum to a random point some distance away. If the new position is sufficiently removed from the valley of the local minimum, then convergence may proceed in a new direction without getting stuck in the same local minimum again. The amount of noise required depends on the local landscape, which is typically unknown to the investigator. Thus, there is some degree of luck involved in getting a network out of a local minimum.

The convergence process of back-error propagation is basically the same as the gradient-descent method, which derives from traditional statistical methodology. Intuitively, gradient descent can be visualized as a skier who is dropped at a randomly selected point on a mountainous terrain. The skier looks about and finds the direction of steepest descent, and then takes an incremental step in that direction. He repeats this process until he reaches a minimum in the terrain. A complete proof of the analogy between back-propagation and gradient descent is given in *Parallel Distributed Processing* (Rumelhart & McClelland 1986, Vol. 1, 323–4).

It should be noted that back-propagation provides more than just the gradient-descent optimization method. The back-propagation network, when successfully trained, finds a way of mapping an arbitrary set of input patterns to an arbitrary set of output patterns. This mapping is found without any knowledge of a mathematical function that may relate the output patterns to the input patterns. Traditional curve-fitting analysis techniques rely on *a priori* knowledge of the form of this mathematical function.

NETWORK TESTING AND PERFORMANCE

Typically an application of back-propagation requires both a training set and a test set. Both the training set and the test set contain input/output pattern pairs. While the training set is used to train the network, the test set is used to assess the performance of the network after training is complete. In a typical application both sets are taken from real data, although sometimes simulated data is used as well. If available data is scarce, then small amounts of noise may be added to the data to simulate additional patterns for the training or test sets. In any case, the training and test sets should use patterns typical of the type of data that the network is to encounter later. To provide the best test of network performance the test set should be different from the training set.

As an example, the network in Figure 4-12 classifies patterns with the same general shapes as those shown in the training set. Patterns with a peak in the middle are of type 1, patterns with a valley in the middle are of type 2, and flat patterns are of type 3. Both the training and test sets include patterns of these three general types, but the same pattern does not appear in both sets.

Figure 4-13 shows an example test set for the network given in Figure 4-12. The test set in this case had 28 input patterns, and included six noisy versions of each of the training patterns. Noise values of ± 0.1 were added to one of the values in each of the input patterns (shown as error bars). The network performed 100% successfully on these test patterns. Additional test patterns were made from horizontal lines of various heights. The network was able to classify the higher lines as type 3, but failed to classify the lower lines because the network failed to produce much activation in any output unit, and thus responded incorrectly.

To put this example in context, suppose that each pattern in the training set from Figure 4-12 is the sales record of a company for a three-month period. A neural network is built to classify three different types of sales patterns—a

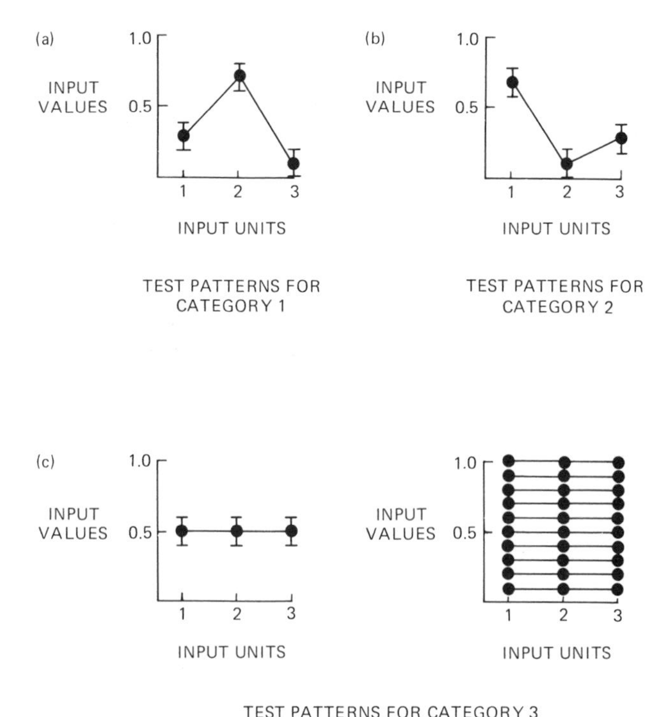

Figure 4-13. A test set for the training set in Figure 4-12.

peak, a valley, and a flat sales record. The neural network is to be used to input sales records and output the type of sales record. If performance is not satisfactory, another experiment may be done using a larger training set to train a second network. A training set that incorporates more of the variation present in the test examples might lead to better performance of the network. A series of similar experiments may be done to improve the network further. A realistic application of sales or financial assessments would probably require more input data and more inputs to the network, but the same general techniques can be employed to train the networks, evaluate their performance, and improve the training results.

FEATURE DETECTORS—AN EXAMPLE

To illustrate how feature detecting units can be developed by back-error propagation, we have constructed an example that is simple enough that the features organized by the network are obviously distinguishing characteristics for the pattern classes. The network to be trained (shown in Figure 4-14)

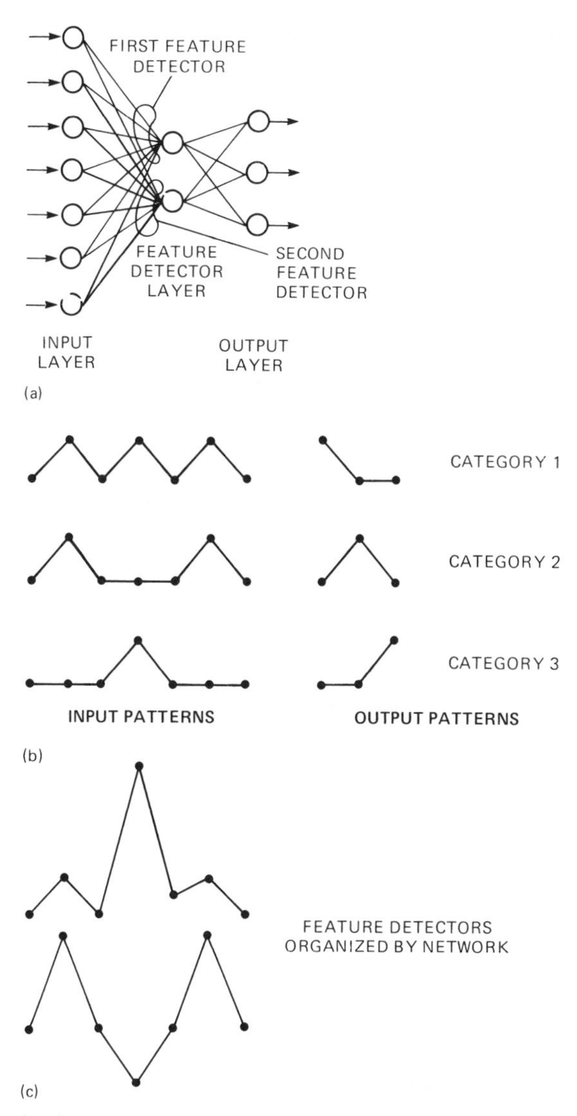

Figure 4-14. Example network training set, and feature detectors.

consists of three layers, with seven processing units in the input layer, two processing units in the hidden layer, and three processing units in the output layer. There are three patterns in the training set (also shown in Figure 4-14), which are to be classified into three categories.

From the combinations of peaks in the training set it is apparent that only two feature detectors are needed. The second and third patterns are combined in the first pattern. Thus, if feature detectors are organized to respond to the second and third patterns, then the first pattern can be identified when both feature detectors are activated.

The network in Figure 4-14 was trained with 460 presentations of the training set. The learning rate was set at 0.6, and the network converged to an RMS value of 0.06. The trained network was able to classify all training patterns correctly.

Features were found by reading weight values from the trained network, as shown at the bottom of Figure 4-14. The weights read were from the first layer of weights, for interconnections that originated at the input layer and terminated at the two hidden "feature-detector" units. Graphs of these weights reflect the features to which each of the hidden units responds. These features roughly match the contours of the second and third training patterns, providing distinguishing characteristics for all three patterns in the training set.

The top layer of the back-propagating network uses the response of the feature detectors in the hidden layer. Each unit in the top layer emphasizes each feature detector according to the value of the interconnecting weight. Thus, the connection from the first feature detector to the third output unit is strong, and the connection from the second feature detector to the second output unit is strong. Both feature detectors influence the first output unit, which stands for a pattern class that has both units' features.

This simple example shows feature detectors that were organized automatically by back-propagation and reflect the salient distinguishing features of the training-set patterns. Although the initial weight values were set to small random numbers, the final weight values "self-organized" to become features that could be used to perform the classification task. The network's only input for this self-organization task was the set of training patterns used during learning.

Although the features organized in this example appeared to the eye to be effective for the classification task, feature detectors do not always organize in a way that is obviously correct to a human observer. For example, in the previous section, the network was trained to classify patterns that looked like a peak, a valley, or a flat graph. The trained network organized feature detectors that appeared like a peak and a valley, but no feature detectors responded explicitly to the third training pattern, a flat graph. The response of the net to the flat graph came about by weighing in many features, none of which appeared flat.

THE EXCLUSIVE-OR: A CLASSIC PROBLEM

The exclusive-or (Xor) function is a classic example for the back-error paradigm. In this problem, a back-error propagating network is trained to perform

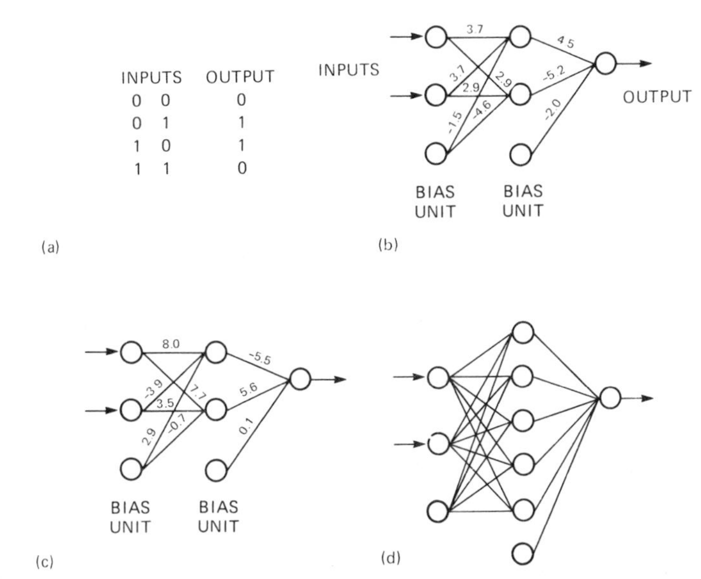

Figure 4-15. The exclusive-or problem: (a) Training set. (b) A two-layer exclusive-or network, trained. (c) An apparent local minimum for the exclusive-or problem. (d) An alternative exclusive-or problem network that is larger and gets caught in local minima less often.

the exclusive-or logic operation shown in Figure 4-15. The perceptron network was unable to solve this problem with a single output unit because the problem is not linearly separable, and its solution requires two layers of adjustable weights.

An Xor network has two inputs and one output. The training set for this problem is simply the Xor function shown in Figure 4-15a. The working network shown in Figure 4-15b is the smallest layered network that accomplishes the Xor function. It is possible to evaluate by hand the summation equations for each of the units to verify that the output is correct.

One of the common problems in doing the Xor problem with a standard back-error paradigm is the presence of local minima. Sometimes when the network in Figure 4-15b is trained, starting with small random weights, the network gets caught in a local minimum. In the case of an apparent local minimum (Figure 4-15c), the network evaluated the first and third entries in the training set correctly, but failed on the second and fourth. Application of more training iterations failed to get better convergence. Although the network adjusted its weights, it failed to converge on the correct answers for all of the patterns in the training set.

Figure 4-15d shows an alternative network topology that trains readily on the exclusive-or problem. This network topology has five units in its hidden layer as opposed to two, and tends to get into local minima less often.

STRENGTHS AND LIMITATIONS OF BACK-ERROR PROPAGATION

The principal strength of back-error propagation is its relatively general pattern-mapping capability; it can learn a tremendous variety of pattern-mapping relationships. It does not require any a priori knowledge of a mathematical function that maps the input patterns to the output patterns; back-propagation merely needs examples of the mapping to be learned. The flexibility of the paradigm is enhanced by the large number of design choices available — choices for the number of layers, interconnections, processing units, the learning constant, and data representations. As a result, back-error propagation might be able to address a broad spectrum of applications.

The largest drawback with back-error propagation appears to be its convergence time. Training sessions can require hundreds or thousands of iterations for relatively simple problems. Realistic applications may have thousands of examples in a training set, and it may take days of computing time (or more) to complete training. Usually this lengthy training needs to be done only during the development of the network, because most applications require a trained network and do not need on-line retraining of the net.

Back-error propagation is susceptible to training failures — the network never converges to a point where it has learned the training set. We have illustrated one such example with the exclusive-or problem, where the network becomes stuck in a local minimum. Additional training does not appear to improve this network.

A variety of special techniques have been developed in an attempt to decrease convergence time and to avoid local minima. A "momentum" term is sometimes used to speed convergence procedures (see Lippman 1987). Parker (1987) summarized a variety of additional techniques that can help the network adapt more efficiently. Improvements in convergence have also been found by varying the learning parameter η by starting with a larger value for η and progressing to smaller values. Techniques for avoiding local minima include changing the network or the training set, and adding noise to the weights. In spite of these improvements, applications developers utilize a variety of specialized accelerator boards, parallel processing machines, and other fast computers in training back-error propagation nets.

It should be noted that back-error propagation was not designed to model biological systems, and does not include many of the biological structures found at synapses and in nerve interconnections. Although biological systems have neurons that perform a type of summation of inputs, and have varying interconnection strengths, a scheme that does error-differencing and back-error propagation has not yet been identified in biological systems.

References

Lippmann, R. P. April 1987. An introduction to computing with neural nets. *IEEE ASSP Magazine,* pp. 4–22.

Parker, D. B. 1982. "Learning Logic," Invention Report S81-64, File 1, Office of Technology Licensing, Stanford University.

Parker, D. B. 1987 "Optimal Algorithms for Adaptive Networks: Second Order Back Propagation, Second Order Direct Propagation, and Second Order Hebbian Learning." Proc. of IEEE the First Intl Conference on Neural Networks. San Diego, CA.

Parker, D. B. 1987. Second order back propagation: Implementing an optimal $O(n)$ approximation to Newton's method as an artificial neural network.

Rumelhart, D. E. and J. L. McClelland. 1986. *Parallel Distributed Processing,* Vols. 1 & 2. Cambridge, Mass.: MIT Press.

Werbos, P. J. 1974. Beyond regression: New tools for prediction and analysis in the behavioral sciences. Thesis, Harvard University.

5

Back-Propagation Applications and Examples

Back-error propagation networks demonstrate a surprising capability for a very broad spectrum of applications, including image classification, speech synthesis, sonar return classifications, knowledge base systems, information encoding, and many other pattern classification and perceptual problems. In this chapter we illustrate the basic techniques used in applying back-error propagation by describing a series of applications studies and examples.

We begin with a description of NETTalk, the neural network designed by Sejnowski and Rosenberg to learn to read English text aloud. The NETTalk study addressed a number of important points that usually arise in applications of back-error propagation: choosing the number of layers, the number of hidden units, and the length of training. Also studied were issues of damage to the trained network and retraining after damage. This application also demonstrates progressive stages in training the network, and illustrates a distributed internal representation in the trained network.

The second example of back-propagation is taken from two-dimensional shape recognition. The actual problem described is a highly simplified classification task motivated by medical image processing. In this task, the network is trained to distinguish four different images by their shape and general visual pattern. A hidden layer of units organized a striking set of feature detectors that can be displayed as two-dimensional patterns. This example illustrates two important aspects of back-propagation: the presence of distinguishing characteristics in the feature detectors and redundancy in the trained network.

The third study described addresses the use of back-propagation for classifying patterns that are presented over time (i.e., not all the pattern is presented at the same time). The time-delay neural network (Waibel 1988;

Waibel et al 1989) employs time-delay components at each layer of the network to utilize a time history of each incoming data signal. In this way the network can be trained to identify patterns that arrive over a period of time. Here we describe how such a network is trained to recognize spoken syllables.

NETTALK

Perhaps the most striking neural network application is the network that learns to read aloud. NETTalk, developed by Sejnowski and Rosenberg (1987) at Johns Hopkins University, learns to translate segments of English text into phonetic notation for pronouncing the text. The phonetic notation that is output can be given to a speech generator and pronounced out loud automatically.

NETTalk is trained using actual examples of English text combined with the phonetic notation. Training NETTalk networks took 1 – 2 weeks of CPU time on a VAX; after training was complete, performance levels were over 90% correct. This performance level is very high considering that NETTalk was the first applications study of its kind. A realistic applications use, however, would require better performance. Future network designs may possibly enable us to attain these higher performance levels.

Figure 5-1 shows the network topology for NETTalk. The network consists of three fully interconnected layers. The input layer has 203 input units, the hidden layer 80, and the output layer 26. The bottom layer inputs English

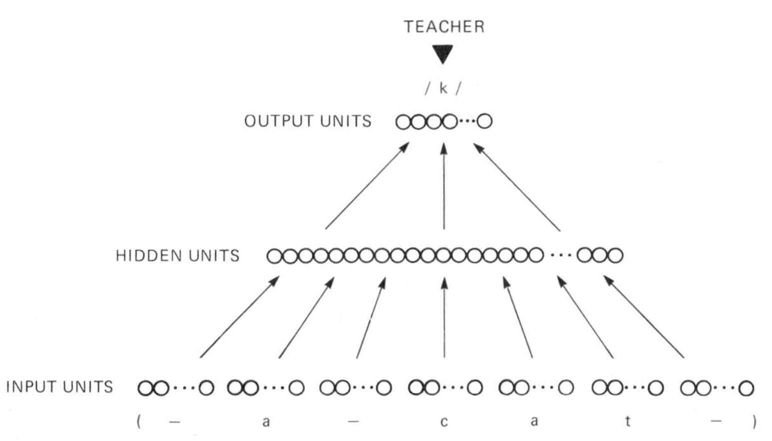

Figure 5-1. The NETTalk back-propagating neural network (from Sejnowski and Rosenberg, 1987. Parallel Networks That Learn to Pronounce English Text, *Complex Systems*).

text, and the top layer is presented with pronunciation notation for sounds to be pronounced from the input text. The NETTalk study illustrates the extent to which pronunciation rules can be mastered by a three-layer back-propagation net with only a single feature detection layer.

Because the English language requires context for recognition of letter pronunciation, an input window that centered on the letter to be pronounced but also included three letters to the right and left of the center letter was used. This input window of seven letters was stepped character-by-character through the English text. Each position was treated as an input pattern in the training set.

The 203 input units were divided into seven groups of 29 units each. Each of the seven groups represented a single character in the text stream and the seven groups together represented the seven-letter window. Each of the 29 units in a group was assigned to represent a single character: the 26 letters of the alphabet plus two punctuation indicators and a word boundary notation. A single unit was activated in each group, corresponding to the letter or character in the text stream. For example, if the first letter in the seven letter window was an A then the first unit in the first group was activated and the other 28 units in the first group were deactivated.

The 26 output units represent different phonetic notation symbols, for individual sounds or stresses: The 21 articulatory features were each represented by a single unit, and five additional units represented specific stresses and syllable boundaries. The output unit with the highest activation level indicated the phonetic symbol the network produced. This symbol was the network's response to the character at the center of the input window, and was presented to a sound generator for pronunciation.

The size of the hidden layer was chosen after experimentation with this back-propagation system. There is an inherent trade-off to be made — more hidden units result in more time required for each iteration of training; fewer hidden units result in a faster update rate but provide fewer feature detectors. NETTalk's hidden layer of 80 units was chosen as a result of learning experiments that indicated that 80 units were sufficient for good performance but not prohibitive in terms of training time.

The experiments done to arrive at the hidden layer size used a range of sizes. Some training experiments used up to 120 hidden units, some none at all. Performance was surprisingly high with no hidden units — up to about 80% correct pronunciation was found. When there were 120 hidden units, the network gave a greater than 90% correct performance. However, there was not much difference between 60 and 120 hidden units — apparently the first 60 feature detectors were the most important to maintain high performance levels.

The network was trained from a database of 1,000 words of transcribed

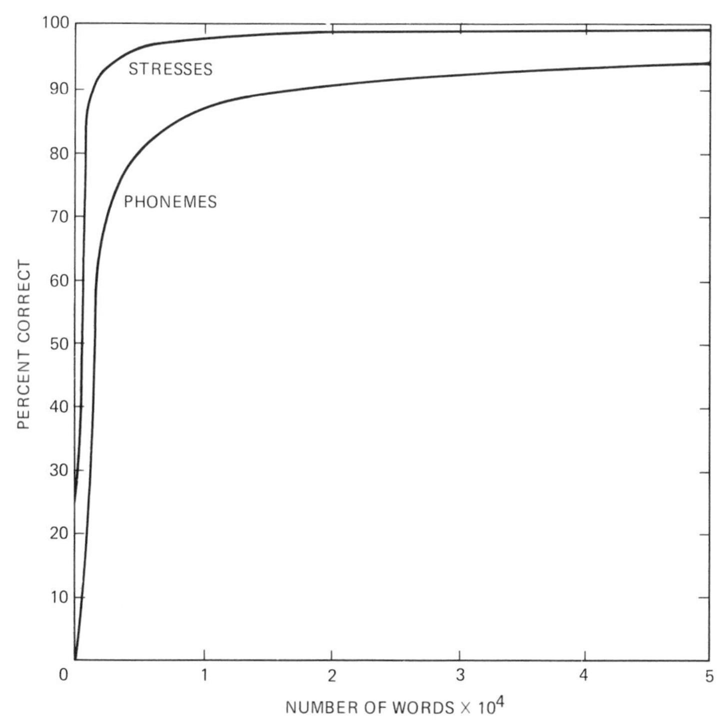

Figure 5-2. Performance of a NETTalk network as a function of the amount of training (from Sejnowski and Rosenberg, 1987. Parallel Networks That Learn to Pronounce English Text, *Complex Systems*).

speech (for details, see Sejnowski & Rosenberg 1987). Figure 5-2 shows the learning curve. The training set was presented up to 50 times to obtain these results. Output units that represent stresses are graphed separately from output units that represent sounds. The stresses were learned very quickly—the network had near perfect performance for stresses after five passes through the training set. Phonemes were learned more slowly—a 95% correct performance for phonemes required 50 presentations of the training set.

The network progressed through different stages during training. First it distinguished between vowels and consonants. At this stage the network predicted the same vowel for all vowels and the same consonant for all consonants. Then, at the next stage, the network recognized boundaries of words, and its pronunciation sounded like pseudowords. By 10 passes through the training set, most of the text was intelligible. These different stages were reminiscent of the way that children learn to speak. A further similarity between the network and people is seen in Figure 5-2, which shows that the

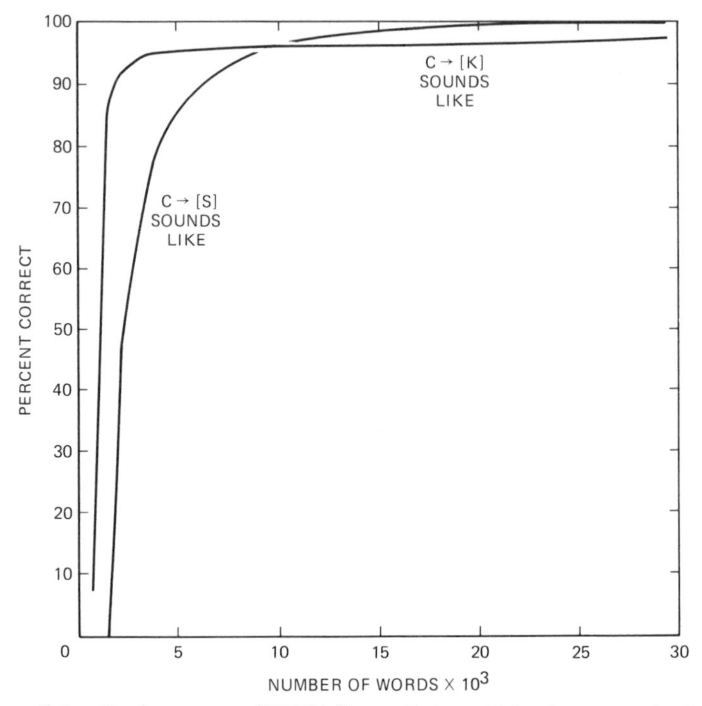

Figure 5-3. Performance of NETTalk on distinguishing between the *k* and *s* sounds for the letter *c*. Percent correct pronunciation is graphed as a function of the amount of training (from Sejnowski and Rosenberg, 1987. Parallel Networks That Learn to Pronounce English Text, *Complex Systems*).

network's learning follows a power law, which is characteristic of human skill learning.

Experiments with NETTalk indicated that there were some phonetic rules that the network never learned. Figure 5-3 shows such an example: Here, the network was required to distinguish between the hard and soft pronunciation of the letter *c*. Figure 5-3 shows the performance level (percent correct) as a function of how many words were presented during training. One curve shows pronunciation performance for words that have a *c* that is pronounced hard (as a *k* sound). The other curve shows pronunciation performance for words that have a *c* that is pronounced soft (as an *s* sound). If the network in Figure 5-1 is trained on more and more words, correct pronunciation of each type of *c* increases. However, a plateau is reached. More than 3,000 words in the training set does not increase the network's performance for pronouncing the letter *c* as *k*. The performance level here stays at about 95%.

Certain rules apparently are not learned by a three-layered back-propaga-

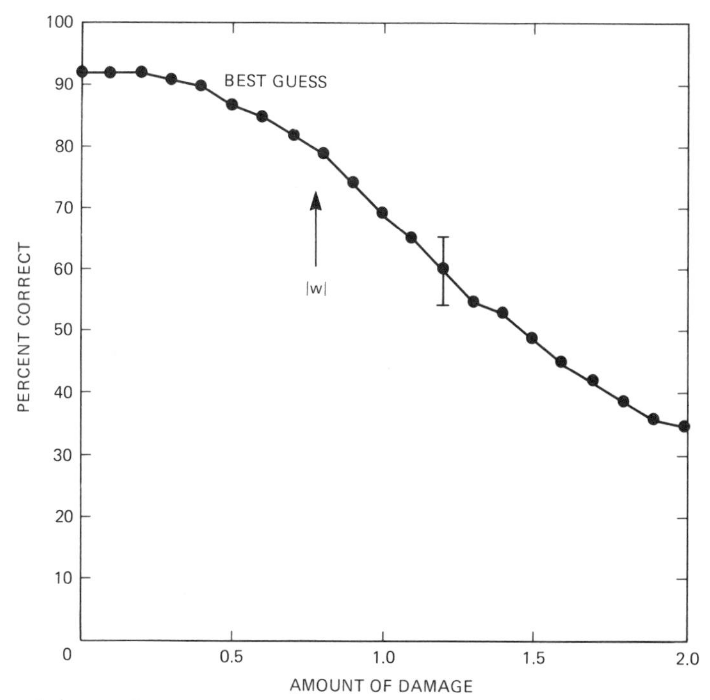

Figure 5-4. Performance of a trained NETTalk network as a function of the amount of damage done to the weights (from Sejnowski and Rosenberg, 1987. Parallel Networks That Learn to Pronounce English Text, *Complex Systems*).

tion network. Increasing the number of examples in the training set does not cause the network to learn the distinction between soft and hard pronunciation of the letter *c* any better.

Additional experiments were done on damage to the network. Damage was done to the network by adding a small random number to each of the weights. The random number was taken from a uniform distribution over intervals from 0.1 to 2.0. Performance of the resulting network was affected by the damage, and deteriorated as the amount of damage grew (see Figure 5-4). The performance was affected only slightly for small amounts of damage; performance dropped off more rapidly with larger amounts of damage. This relationship shows that network performance is resilient to small amounts of damage.

A further indication of the robustness of the network in the face of damage is shown in Figure 5-5, which shows that if the damage is not too severe, the network can recover readily with more training. The bottom trace shows the curve for original learning, giving performance (percent correct) as a function of the number of words presented during training. The top trace shows relearning after damage; damage in this case was done by adding random values between −1.2 and 1.2 to all weights. Relearning after damage occurred faster than the original learning.

Networks with four layers were also trained in the NETTalk study. One four-layered network used the input and output layers described previously in addition to two hidden layers with 80 units each. Performance was then compared to the three-layered network with 120 hidden units; the total number of weights in these two networks differed by only 10%. Both networks did about the same in absolute performance. The four-layer network, however, was able to generalize better—it pronounced new words better than the three-layer network with 120 hidden units.

The NETTalk network organized feature detectors that tended to respond to particular sounds or groups of sounds. The patterns that individual feature detectors responded to were not obvious in many cases; however, the hidden layer taken as a whole had a distributed representation of the key features of the data. An interesting point about this organization arose when the same network was trained twice on the same data, each time starting at different randomized weight values. The resulting networks could have the same performance and learning characteristics on a particular task, but differ com-

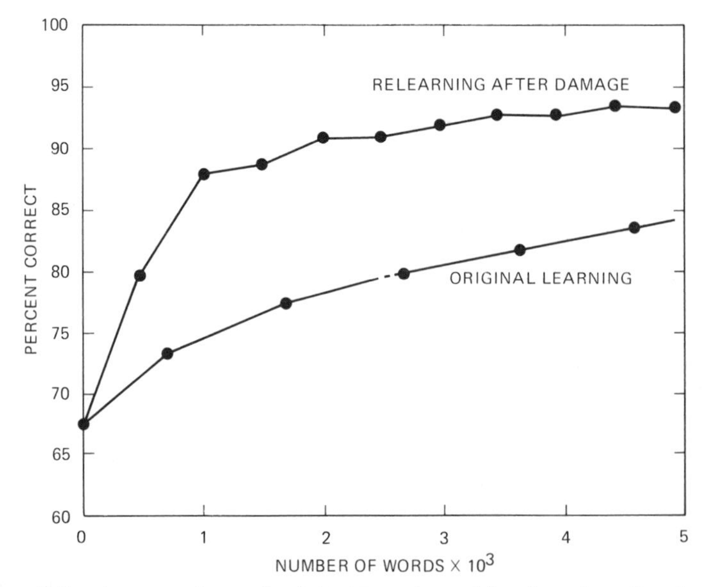

Figure 5-5. A comparison of original learning with relearning after damage. Performance is shown as a function of the amount of training (from Sejnowski and Rosenberg, 1987. Parallel Networks That Learn to Pronounce English Text, *Complex Systems*).

pletely at the levels of synaptic strengths and single-unit responses. Hence, the details of the feature detectors differed even though the performance of the networks were the same.

NETTalk demonstrates that most of the expertise needed to read English text aloud can be learned by a three-layered back-propagation system. It is humbling to observe that such a simple network can actually learn a task that is so fundamental to human existence. The NETTalk system also illustrates a number of important considerations in back-propagation studies. Experiments dealt with hidden layer sizes, training sets, learning curves, network damage, relearning after damage, and the particular organization of the feature detectors. These issues arise in other back-propagation applications as well.

TWO-DIMENSIONAL FEATURE ORGANIZATION

Systems that can identify and classify two-dimensional images based on their shapes and other visual characteristics are very much needed in applications such as the identification of handwritten numerals (e.g., on checks), reading of handwritten characters, sorting of parts in industrial production, automated defect inspection, and medical image processing. Although these applications are not expected to be solved entirely by back-propagating systems, neural networks are likely to be a key component in their solution.

Two-dimensional patterns provide us with a visually intuitive way to show how back-propagating systems work internally. The feature organization can readily be seen by eye in a two-dimensional image classification task. Here we show a simple example in which a back-propagating neural network (Wasserman 1988)[1] organizes a set of two-dimensional features when it learns to classify a collection of different shapes (Dayhoff & Dayhoff 1988). This shape classification problem was motivated by medical image classification tasks that have the potential to benefit from the development of automated classification systems.

Figure 5-6 diagrams the three-layered network used in this study. The layers are fully interconnected. Both the input and output layers are arranged as two-dimensional grids (7×9), and the hidden layer in this case has 16 processing units.

Figure 5-7 shows the training set adapted from a medical image processing task. A coarse-grid representation of four images was made to simplify the

[1] Thanks to Phil Wasserman for making his software available.

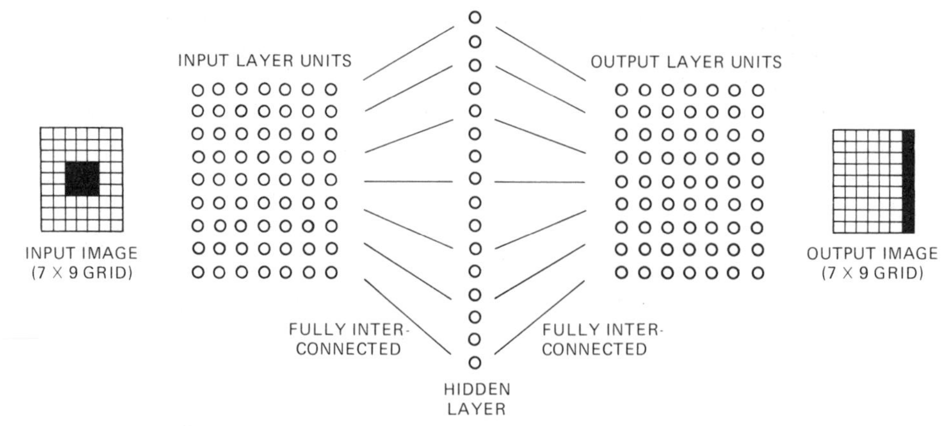

Figure 5-6. A three-layered back-propagation network used to perform a simplified medical image classification task (from Dayhoff and Dayhoff, 1988, © IEEE).

pattern recognition task. The four images were based on the shapes of different one-celled organisms. The organisms were amoeboid cells, distinguishable by the thickness of their membrane and the size and organization of their nucleus. The coarse-grid representation is shown in Figure 5-7 along with the original images of the amoebae. (Note that the training set is a considerable simplification of real-life images.)

The network in this study was trained to output a grid with a vertical slab of activity. The slab is to appear at a different place based on the network's classification of cell type. Four different positions are shown, corresponding to the four different cell types. These are the target outputs of the neural network.

The training set was presented to the network 3,000 times. The resulting network was able to classify all pictures in the training set, and in addition showed good performance on a test set with noisy pictures.

The feature detectors that were organized by the network during training are shown in Figure 5-8a. Figure 5-8b illustrates how these feature detectors were drawn based on the trained network. Each feature detector corresponds to a single unit in the hidden layer of the network. The incoming connection weights for each hidden unit make up a 7×9 grid. The values of these weights were quantized and drawn in a grid. Black boxes represent inhibitory connections, striped boxes represent positive connections, and cross-hatched boxes were strongly positive. White boxes represent weights with very low absolute values. Feature detectors developed emphasized the following areas of the images: outer edge, inner edge, central square, four corners of central square, "+" inside central square, and background.

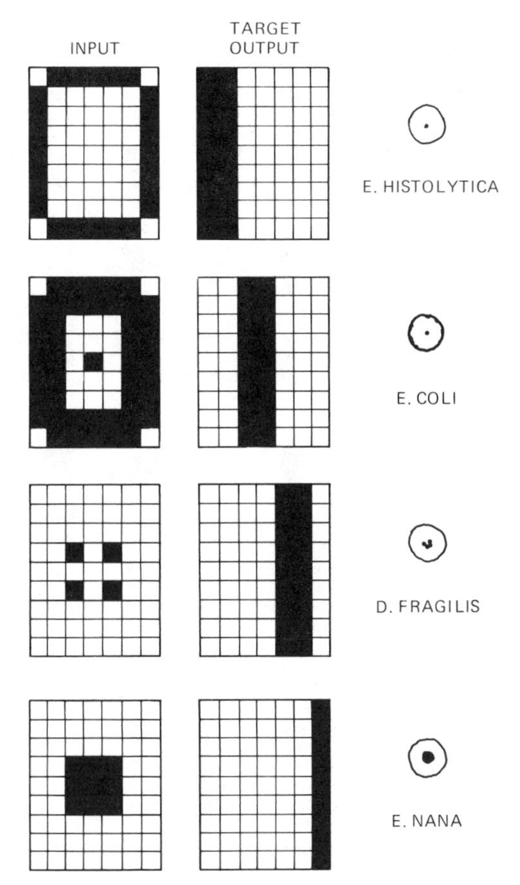

Figure 5-7. Four different types of ameboid cells with their coarse-grid representations. The target output patterns for the network are also shown (from Dayhoff and Dayhoff, 1988, © IEEE).

Each feature detector illustrates the incoming pattern that the hidden unit responds to best. Thus, an input pattern that is exactly the same as a given feature detector will activate the corresponding feature-detecting unit best. If the input pattern matches a subset of the feature, then the corresponding unit will be activated but not at maximum activation. An input pattern that does not match the feature will not activate the corresponding hidden unit significantly.

The feature detectors that were organized (Figure 5-8a) are similar to fragments of the images in the training set. It is possible to pick out by eye some of the features that are distinguishing characteristics of two or more images. For example, the presence of a thin cell boundary appears in features

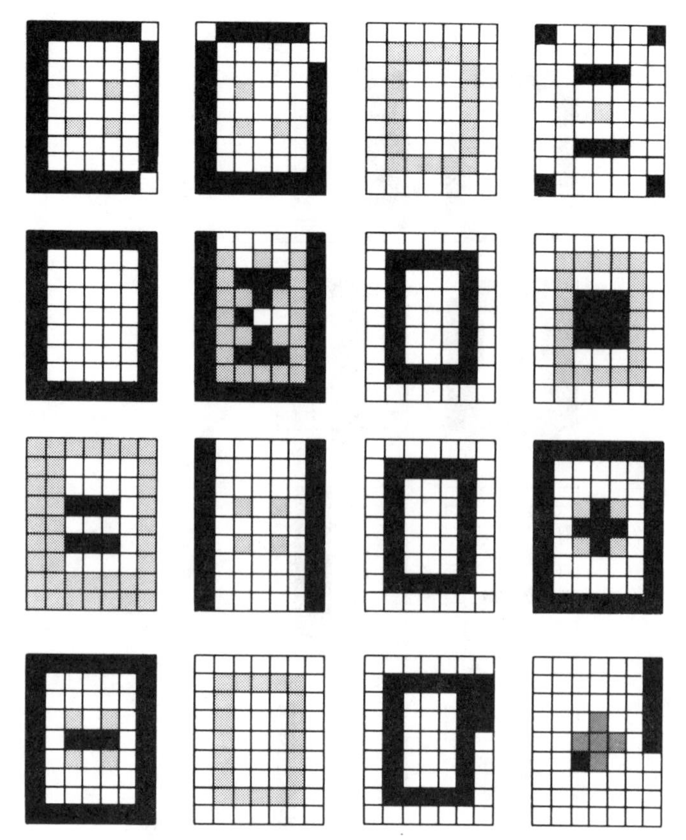

Figure 5-8. (a) Sixteen feature detectors organized by the network shown in Figure 5-6 to solve the classification task shown in Figure 5-7 (from Dayhoff and Dayhoff, 1988, © IEEE).

1-2, 5-6, and 12-13. An inner cell boundary elicits a response from features 7, 11, and 15. The nuclei in the last two training patterns are distinguished by the + pattern in feature 16, and can be partially distinguished by the pattern of four dots in features 1, 10, and 13.

The features organized in this example are highly redundant—some are almost exactly the same (e.g., 7 and 11). Many pairs of feature detectors contain similar components. This redundancy is a natural part of back-error propagation training because redundancy is organized automatically when an excess of hidden units is supplied. Furthermore, the redundancy is usually distributed throughout the hidden unit layer, as in Figure 5-8a. Redundancy in feature detectors contributes to the capabilities of the network by helping to make the network performance resilient to errors or failures in individual units or weights.

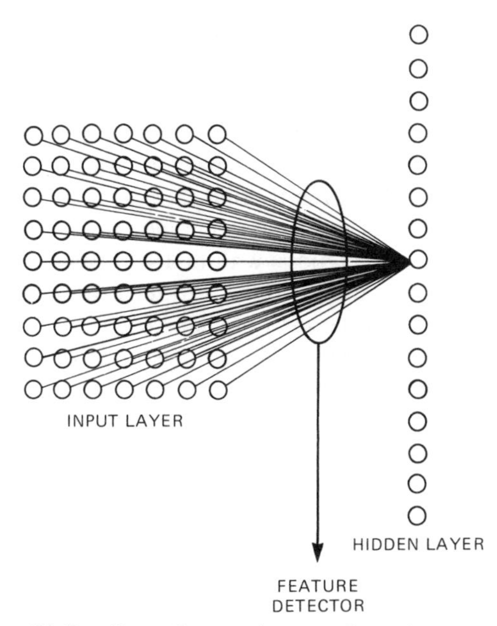

Figure 5-8. (b) Reading a feature detector from the trained network.

If the applications goal is to include redundancy in the system, the redundancy makes the system more tolerant to damage of individual units or weights. Computing systems that are tolerant to damage of components are needed, especially in applications of military or medical systems. If the applications goal is to construct the smallest system possible, then the number of hidden units can be decreased and performance would continue to be good. The example shown here can be decreased to 12 hidden units without significant degradation of performance.

The way that these feature detectors are used by the output layer is specified by the second layer of weights. Weights from the hidden layer to the output layer indicate how strongly each feature is weighed by the network. Feature 16, for example, is strongly weighted in its connection to the third and fourth slab of output units, and is used to distinguish these images.

THE TIME-DELAY NEURAL NETWORK

One of the chief limitations of traditional neural network architectures is in dealing with patterns that may vary over time and that require a period of time to be presented. Standard neural network architectures such as back-error propagation require an input pattern to be presented all at the same time and have no explicit architecture built into the network to deal with patterns such as speech, which are presented in parts over a period of time, with a lag between when the pattern begins and when it ends. A spoken word is presented to the listener over a period of time, and a spoken sentence is presented over a longer period of time. The human nervous system can recognized these patterns almost immediately even though they are not presented all at once.

The basic architecture of the time-delay neural network (TDNN), as described by Waibel and colleagues (Waibel 1988; Waibel et al 1989) has a variety of time-delay components built into its structure so that it can deal with temporally presented patterns. Back-error propagation is used as the learning algorithm. Considerable success has been obtained with this network in the area of speech-recognition studies.

Figure 5-9 shows the basic unit of the TDNN. A series of input channels

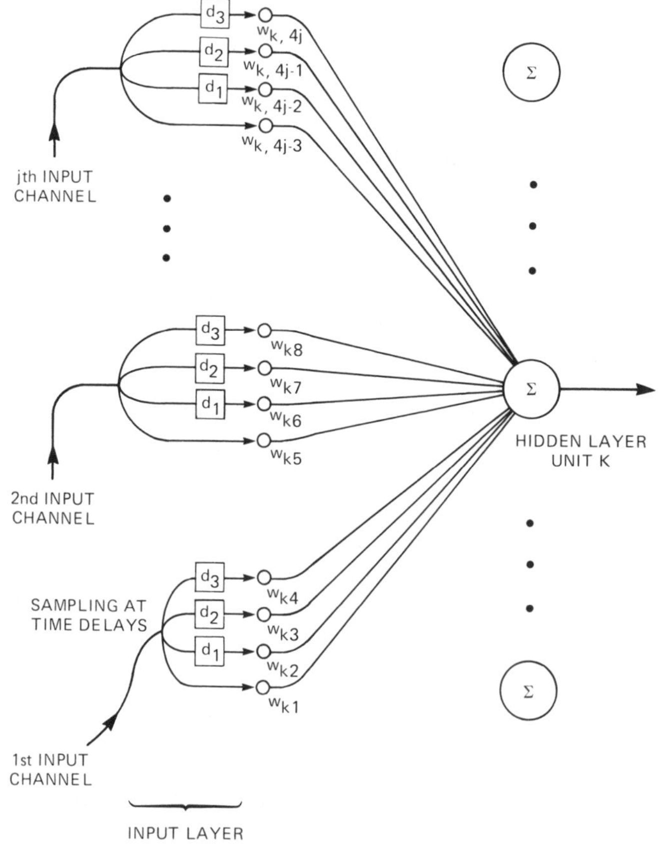

Figure 5-9. A processing unit from the time-delay neural network.

provides signals for the network. Each channel is sampled over a period of time, and past values are input to the network after delays. The bottom input for each channel is the present value of the channel signal; the other inputs are previous values, each delayed by a fixed amount. In principle, one can include any number of channels, any number of time delays, and any collection of time delay values in such a unit.

The input units and delays are considered the first layer of the network. The second layer is a hidden layer, and the unit shown at right of Figure 5-9 is a hidden layer unit. This unit produces a sequence of outputs, which are stored over a period of time. Then the present output together with delayed outputs from the past are fed into the next layer. The TDNN can employ two or more layers of time-delay units like the one in Figure 5-9.

Each interconnection of the network has an associated weight. The weight is adjusted according to the standard back-error propagation paradigm. Temporal relationships, then, become encoded in the network by means of the values of weights associated with the different time delays. For example, suppose that the weight for a particular interconnection has an unusually high positive value. Observe the time delay associated with the unit where the interconnection originates—it is emphasized by the high weight value and has more influence on the output of the network than other time delays.

In a series of initial experiments done with a time-delay neural network, a four-layer network was trained to recognize and classify a set of spoken syllables. The incoming speech was split into 16 separate frequency bands and each band was assigned to a separate input channel. The input signal for each channel was the amplitude for a narrow band of frequency as a function of time. Three input units were assigned to each channel. The first unit received the input signal itself and the other two units received the input signal delayed in time, with a different time delay for each unit.

The trained network was able to classify three different consonant sounds: *D, B,* and *G.* A single unit in the output layer was taught to respond to each consonant. The neural network was trained to recognize the consonant sounds regardless of the vowel sound that followed, and a variety of different vowel sounds were used.

Figure 5-10a shows the response of the trained network to the activation from the syllable *DA.* For comparison, Figure 5-10b shows the activation pattern resulting from input of the syllable *DO.* Although the activation patterns in the first and second layers differ considerably, the third layer has a similar activation pattern for both. These cause the output layer of the network to respond with the classification *D.*

The TDNN performed successfully, and its performance was better than a leading nonneural network approach, which utilizes hidden Markov models. Furthermore, the researchers have continued their work to obtain even better

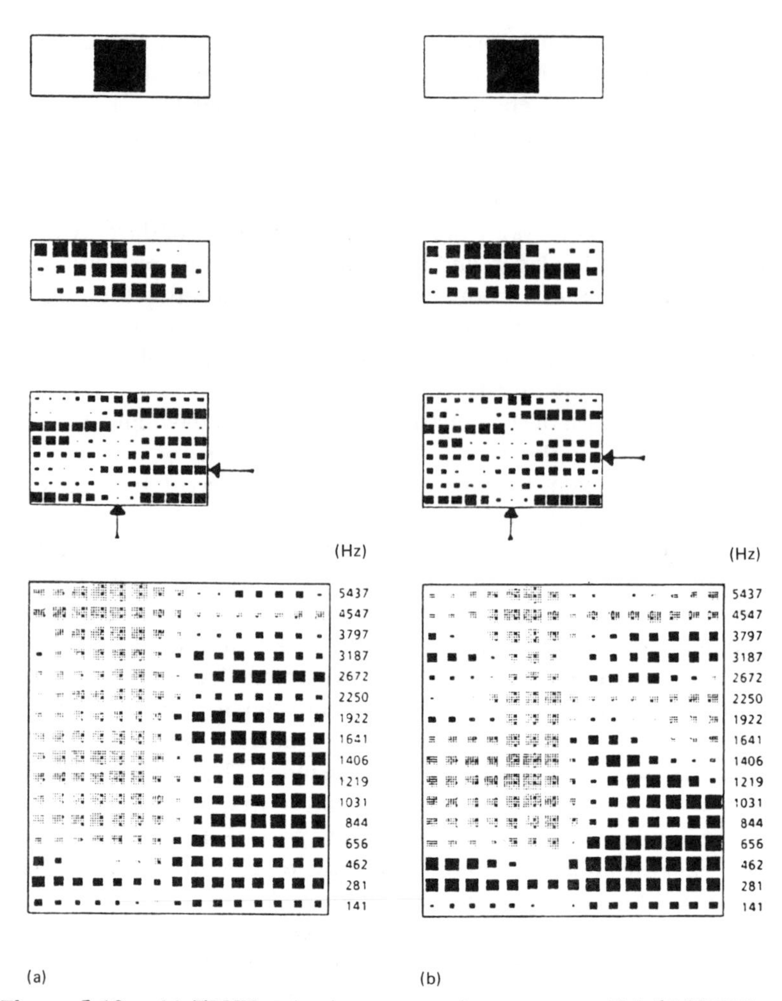

Figure 5-10. (a) TDNN activation patterns in response to *DA* (b) TDNN activation patterns in response to *DO* (from Waibel et al. Phoneme recognition using time delay neural networks, © 1989, IEEE *Trans. on Acoustics, Speech, and Signal Processing*).

performance with a series of similar subnets that divide the classification task (Waibel 1988). These research studies have made striking progress toward designing a system that can dynamically recognize speech, a very difficult task.

References

Sejnowski, T. and C. Rosenberg. 1987. Parallel networks that learn to pronounce English text. *Complex Systems* 1(1):145–68.

Dayhoff, R. E. and J. E. Dayhoff. 1988. Neural networks for medical image processing. IEEE SCAMC Proceedings, pp. 271–75.

Waibel, A.; T. Hanazawa, G. Hinton, K. Schikano, and K. Lang. 1989. Phoneme recognition using time delay neural networks. *IEEE Trans. on Acoustics, Speech and Signal Processing* **37**, March 1989.

Waibel, A. 1988. Consonant recognition by modular construction of large phonemic time-delay neural networks. Neural Information Processing Systems, Morgan Kaufman Publishers, San Mateo, Calif., pp. 215–23.

6

Competitive Learning and Lateral Inhibition

In this chapter we cover networks in which processing units act—through competition and inhibition—in opposition to one another. The first such neural circuit presented here is the *competitive learning* paradigm, which uses a layer of processing units that compete with one another, resulting in a network that can be used for pattern-classification applications. The second type of circuit is *lateral inhibition,* which uses a processing unit layer in which each unit inhibits other nearby units. The resulting network is useful for preprocessing of sensory data, to accomplish edge enhancement and contrast enhancement.

Competitive learning is the first paradigm we have discussed in this volume that performs unsupervised learning. In unsupervised learning, the network is presented with a set of training patterns, but is not given a target answer (usually a classification category) for each input pattern; the network organizes the training patterns into a set of classes on its own. This extends the capabilities of neural networks to pattern-classification applications where the target classifications are not known a priori but the data can nevertheless be sorted into different categories.

The basic competitive learning network consists of two layers—an input layer and a competitive layer. In the competitive layer, the units compete for the opportunity to respond to the input pattern. The winner represents the classification category for the input pattern. The competition may be accomplished by means of an algorithm that assigns the winning unit, or, alternatively, through inhibition among the units in the competitive layer. In the case of inhibition, the competitive layer progresses to a state in which only the winning unit is active.

In contrast, the lateral inhibition circuit is represented with only a single layer of processing units, and each unit inhibits only those units that are nearby. The result is that the output of the network enhances some of the edges and transition areas that appear in the input patterns.

Competitive layers and inhibitory connections are key elements to a number of neural network paradigms. For example, the Kohonen feature map, discussed in Chapter 9, employs a competitive learning scheme along with other structures; the counterpropagation paradigm, covered in Chapter 10, employs a competitive learning scheme in one of its layers. Networks such as ART (adaptive resonance theory) also contain a substructure that is similar to the basic competitive learning scheme (Grossberg, 1988; Carpenter and Grossberg, 1988; Stork, 1989).

COMPETITIVE LEARNING ARCHITECTURE

The first layer of a basic two-layered competitive learning network—the input layer—is made up of processing units that receive input patterns. The second layer is the competitive layer, which classifies the input pattern. The two layers are fully interconnected; each unit in the input layer sends an interconnection to every unit in the competitive layer (see Figure 6-1). There are many ways of implementing competitive learning; we have chosen to describe one of the simplest forms possible for a competitive learning scheme (Rumelhart & Zipser 1986).

Each interconnection has an associated weight. In this scheme, the weights are limited to values between 0 and 1. To simplify the network, we utilize the restriction that the sum of weights to a given processing unit is always 1:

$$\sum_i w_{ji} = 1 \tag{6-1}$$

The input vector is a binary vector of 0/1 values. These restrictions are not necessarily required in more advanced versions of competitive layers (see Chapters 9 and 10). Initially the weights are set to small randomized values that satisfy (6-1).

Figure 6-2 illustrates the basic processing unit in the competitive layer. Its processing takes place in two steps: first the weighted sum is computed and then the competition occurs. The weighted sum for unit j is:

$$S_j = \sum_i w_{ji} x_i \tag{6-2}$$

where w_{ji} = weight to unit j (competitive layer) from unit i (input layer), x_i = activation level (output) for unit i in layer 1, taken from the input vector,

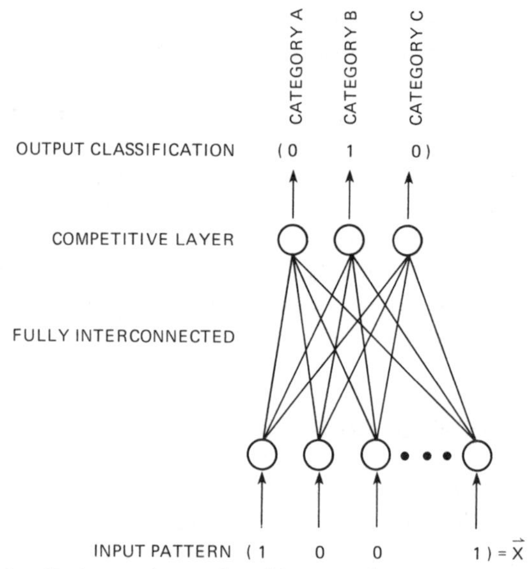

Figure 6-1. Basic two-layered architecture for competitive learning.

and the sum is taken over all units in the input layer. This weighted sum is calculated for all processing units in the competitive layer before the next step takes place. Once the incoming sum is known for each unit in the second layer, the competition among them begins.

In the "winner takes all" scheme, the unit with the highest weighted sum in

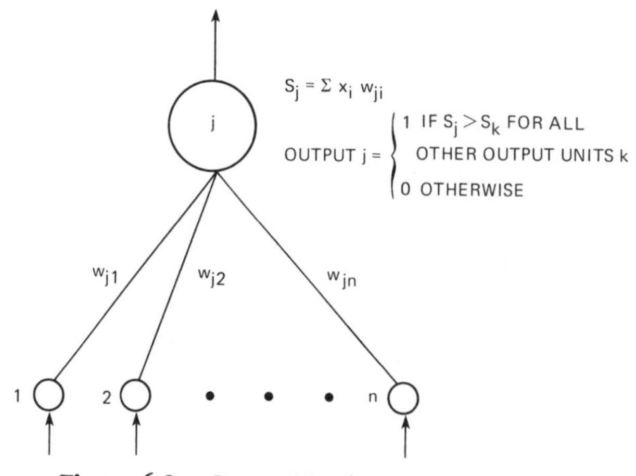

Figure 6-2. Competitive layer processing unit.

layer 2 is assigned as the "winner." This unit is then given a new activation value of 1.0, and all the other units are given the value 0.0. Mathematically,

$$a_j = \begin{cases} 1 \text{ if } S_j > S_i & \text{for all } i,\ i \neq j \\ 0 \text{ otherwise} \end{cases} \tag{6-3}$$

where S_j is the weighted sum from (6-2) and a_j is the activation value for unit j after competition is complete. In the event of a tie ($S_j = S_i$), then, by convention, the unit on the left is selected.

The weights are updated after the "winner" is assigned. Only the weights that go to the winning unit are changed. They are updated according to the rule:

$$\Delta w_{ji} = g\left(\frac{x_i}{m} - w_{ji}\right) \tag{6-4}$$

where g = the learning parameter ($0 < g \ll 1$) chosen by the user, m = the number of units in the input layer that have activation levels of 1.0, and x_i = value of input unit i (0 or 1).

The value of g is set to a small constant and reflects the amount of adjustment to the weights on each iteration. Typical values for g are 0.01 – 0.3. The variable m counts the number of active input units. Note that the weights are initially set to small randomized values that sum to 1.0, as in Eq. (6-1).

The first term in (6-4) causes the weight to be incremented when the corresponding input unit is active (activation value 1), and the weight will be decremented when the corresponding input unit is inactive (activation value 0). The ith weight, for example, will become larger when the input value x_i is 1 and will become smaller when the value of x_i is 0. Note that the weights are between 0 and 1, and the weight becomes closer to the corresponding input value upon updating. This change is illustrated in Figure 6-3, where weights from the first, fifth, and sixth input units are increased because those units are active, while other weights are decreased.

The adjustment in (6-4) causes the weighted sum for the input pattern to be slightly higher if the same input pattern is presented again immediately. In addition, input patterns similar to the current input will also yield a higher weighted sum for the winning unit. Hence, the winner will probably win again when the same or similar patterns are presented afterward. Note that additional weight adjustments may be made in response to different patterns before the same pattern or a similar one is presented again, and these changes could again change the response of the network.

The second term in (6-4) ensures that the restriction in (6-1) is upheld. The sum of the weights going to the winning unit must equal 1. The adjustment

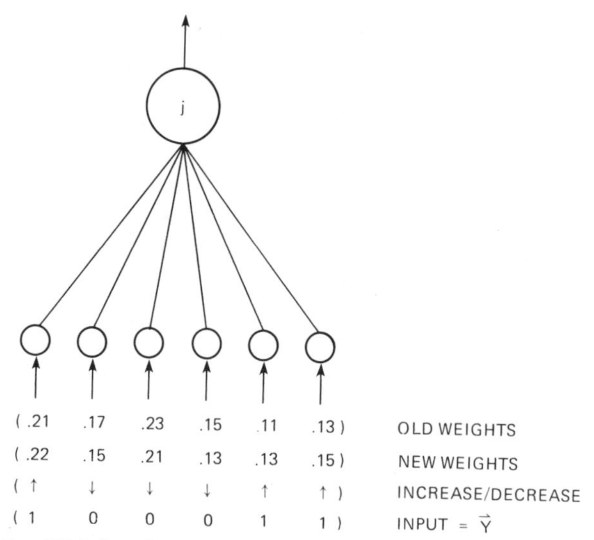

Figure 6-3. Weight adjustments in competitive learning. Weights before and after the input vector **y** is presented.

procedure takes away a small amount of this sum from each weight and adds back amounts that are proportional to the input pattern. The net result is that the sum of the weights, Eq. (6-1), is still the same. Summing (6-4) over all incoming weight changes shows that the total change is 0:

$$\sum_i \Delta w_{ji} = g\left(\underbrace{\frac{1}{m}\underbrace{\sum_i x_i}_{m}}_{1} - \underbrace{\sum_i w_{ji}}_{1}\right) \tag{6-5}$$

$$= g(1 - 1) = 0$$

Example: Classifying Groups of Simple Patterns

In our example, a competitive learning network is trained to classify two groups of simple patterns. The patterns are binary vectors with three entries (Figure 6-4a). The two-layer network used in this classification task is shown in Figure 6-4b. The output layer has two units; the intention of the training experiment is to divide the input vectors into two classes. During training, the network will choose how to subdivide the input patterns into different classes.

The table in Figure 6-4c analyzes the similarities among the patterns in the training set by showing the number of nonmatching entries between each pattern pair. The first two patterns are similar to each other and the last two patterns are similar to each other; a larger difference is found when comparing either of the first two patterns to either of the last two patterns. Thus, the patterns naturally group themselves into two classes.

Figure 6-4d shows the classification obtained by the network after training was completed. Two patterns appear in each class. Patterns in the same class

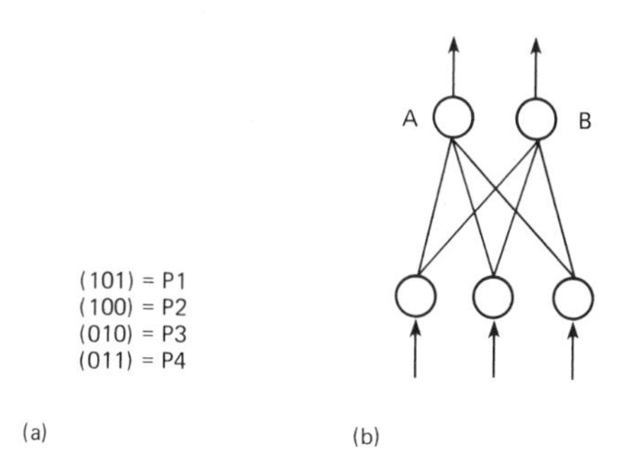

HAMMING DISTANCES:

	P1	P2	P3	P4
P1	0	1	3	2
P2	1	0	2	3
P3	3	2	0	1
P4	2	3	1	0

(c)

P1 = (101) } CLASS A
P2 = (100) }

P3 = (010) } CLASS B
P4 = (011) }

(d)

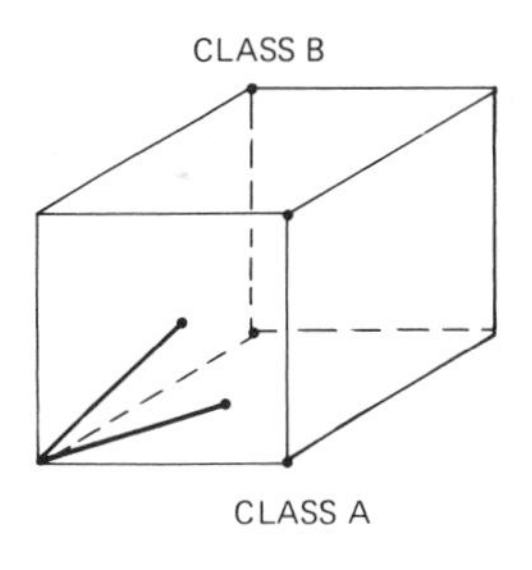

Figure 6-4. A simplified pattern classification example. (a) Training set. (b) Network configuration. (c) Number of nonmatching entries for each pattern pair. (d) Network's classification of patterns. (e) The training set vectors graphed in three dimensions along with the weight vectors of the trained network (shown as points). Weight vectors are $\mathbf{w}_1 = (w_{11}, w_{12}, w_{13})$ and $\mathbf{w}_2 = (w_{21}, w_{22}, w_{23})$.

have only one nonmatching entry, whereas patterns in different classes have two or three nonmatching entries. Thus, the network's classification of the patterns follows the natural categories arising from Figure 6-4c. Repetition of the training experiment using different initial weight values usually resulted in a network that learned to group the patterns correctly.

Figure 6-4e illustrates the weight vectors after training was completed, as well as the vectors in the training set. The corners of the cube plot all possible binary vectors with three values. The weight vector for unit A is nearest to the two vectors in class A. As a result, unit A responds when class A vectors are input. The weight vector for unit B is nearest to the two vectors from class B; unit B wins for class B inputs. The weight vectors are relatively short in this diagram because their components must sum to 1.

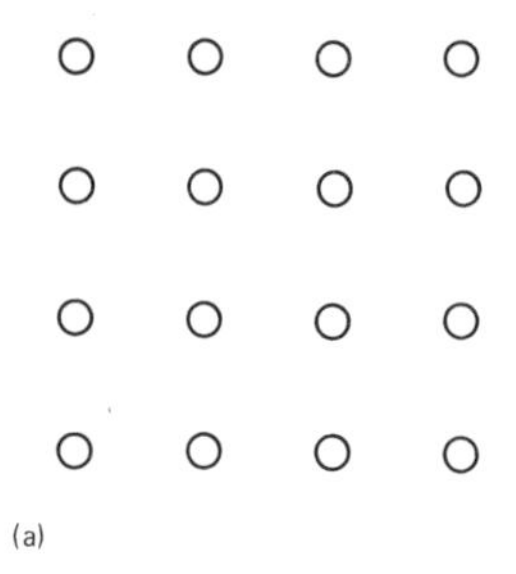

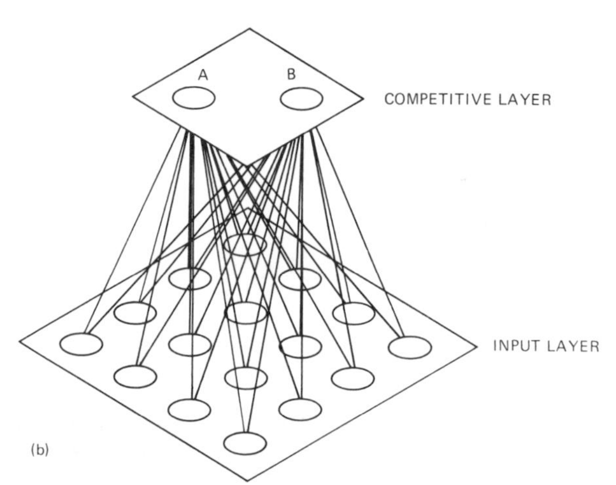

Figure 6-5. A simple competitive learning example. (a) The input layer grid. (b) The network topology.

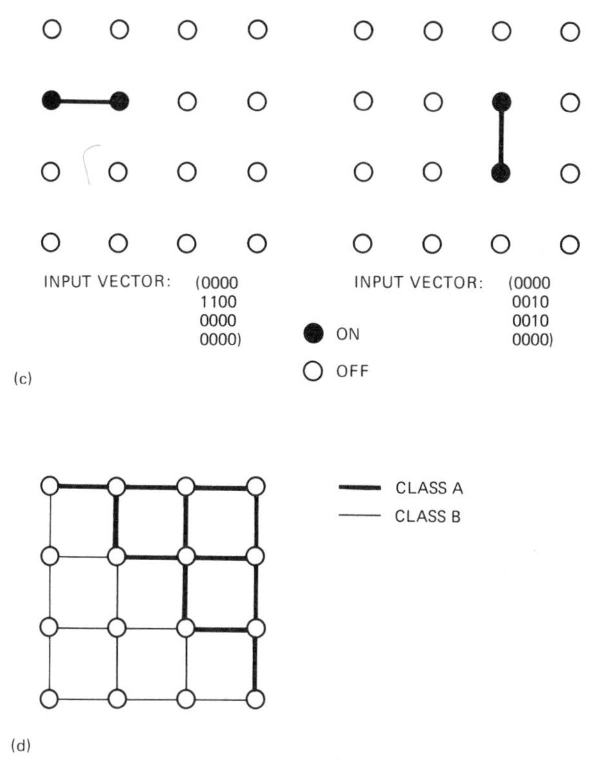

Figure 6-5 (cont.). (c) Two of the training patterns. The training set consisted of all possible pairs of adjacent points. (d) Resulting classification by trained network: dark bars show class A, light bars show class B.

Example: Classifying Regions of a Plane

In this example of a network classifying regions of a plane, points are arranged in a grid, and points that are close to one another tend to get classified into the same category. The network also tends to balance the size of each classification category; this balancing effect can be seen by observing the size of the regions in each category.

The input layer to the network, Figure 6-5a, consists of a 4×4 grid of "off/on" units. There are two units in the top layer, indicating that the network will have two categories for classification (Figure 6-5b). The training set consists of patterns that have two adjacent units on, as shown in Figure 6-5c. All such adjacent pairs are used for the training set.

Figure 6-5d shows one of the common classifications achieved by this type of network. After 1000 iterations with $g = 0.05$, the network divided the input grid in half—input pairs at the top right were in one class and input pairs from the bottom left were in a second class. It was necessary to use small values of g and many iterations to obtain good results.

When the training experiment was repeated with different initial weight values the network did not always reach the classification shown in Figure 6-5d. Sometimes the grid was divided roughly in half with different dividing lines (Rumelhart & Zipser). Since all possible pairs of adjacent units on a full grid were included as training inputs, there was no natural division among the points on the grid, and the network found alternate ways of grouping the inputs.

In a slightly more complicated example, Figure 6-6a, the grid represents the input layer of units in a competitive learning network. There are 24 input units, and the grid is missing two horizontal bars. The output layer used here has only two units. The training set again consisted of pairs of connected units, similar to Figure 6-5c. All pairs of adjacent units were activated for the training set, except that the pairs corresponding to the two missing horizontal bars were not included.

The classification of the input grid after training is shown in Figure 6-6a. The left side was classified into one category, and the right side was classified into a second category. In this example, there was a natural grouping between the two sides of the grid due to the lack of the two horizontal bars in the middle column. As a result, the categories in Figure 6-6a are divided at the place where the fewest interconnections exist.

When a training pattern P is presented, two adjacent input units are activated, and the weights going from those units to the winning output unit are incremented (Figure 6-6b). This increment makes it more likely that the same output unit will win for the next training pattern that includes either of the units activated in P (see Figure 6-6c). The final result is that the same output unit tends to win in response to training patterns with active units from the same region of the grid. More connections in the grid among the nodes in the same region tend to help the network group that region together. The final classification result in Figure 6-6a is thus not surprising.

In this example solution to a graph partitioning problem, the task is to divide a graph into two parts with about half of the nodes in each part while minimizing the number of links that connect nodes from the two different classes. The nodes in each class have more interconnections with other nodes in their class than with nodes in the other class. Pairs of interconnected nodes were activated during training; the result was that each unit in the output layer developed weight vectors that were high for a different side of the grid. Thus, each output unit became activated in response to units on a specific side of the grid.

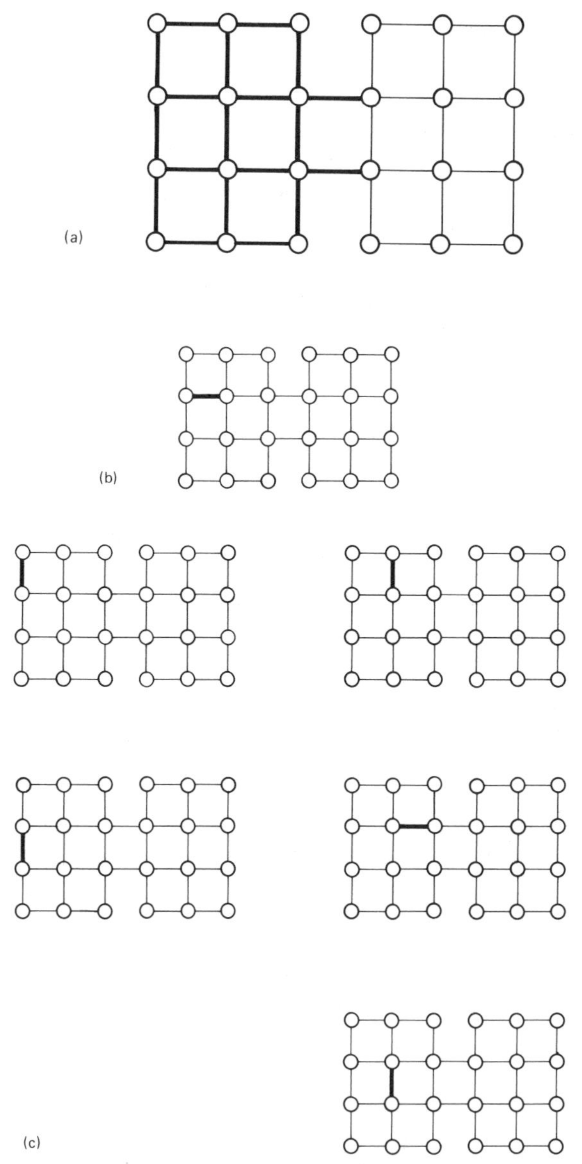

Figure 6-6. (a) Classification of a two-dimensional irregular grid by a competitive learning network. Dark bars show class A, light bars show class B. (b) An example training pattern P. (c) After pattern P is presented to the network, and the weights are adjusted, these five patterns are more likely to yield the same competitive winner in the second layer.

COMPETITIVE LEARNING CHARACTERISTICS

In competitive learning, top-layer units compete for the "right to respond" to an input pattern, and each unit learns to specialize on a group of similar patterns. Thus each output unit becomes a detector for a different class of patterns. The output unit that responds to a pattern identifies the pattern class that the network chooses.

A competitive learning system tends to find clusters of input patterns if clusters in the training set are sufficiently distinct. Thus, when the set of possible input vectors consists of groups of similar patterns, a competitive learning system can learn a set of classifications, and the trained network will be stable in that it will always respond to the same input pattern with the same output unit.

If the input patterns are not well-organized, then the classifications may be unstable, allowing a given input to be responded to first by one unit and then by another. The presentation of each new pattern will cause the network weights to be modified so that the network will continue to respond differently.

In the overall competitive learning scheme, the network develops pattern clusters that minimize within-cluster distance and maximize between-cluster distance (as an approximation). The system also tends to balance the number of patterns captured in each cluster, especially when input patterns vary uniformly. These goals are not rigorously achieved because the network reaches a trade-off.

The groupings developed by a competitive learning system sometimes depend on the starting values of the weights and the order of the input patterns. Consistent performance can be obtained for data that are well-organized into different types of patterns. Choices for learning parameter values and the number of training iterations are usually adjusted experimentally to obtain the best training results.

A number of performance limitations occur with competitive learning. A partially trained network may lower its performance when presented with a new input vector that is quite different from previous vectors because the system sometimes changes its weights in a nonproductive fashion. Competitive learning is not tolerant of pattern translation, and its classifications are not size- and rotation-invariant. These last limitations are not surprising, as the network has no explicit structure to address translation, size, and rotation invariance.

The principal strength of competitive learning is providing a simple pattern classifier that is trained through unsupervised learning. Competitive learning layers can also be included in other more complex networks. A larger net can be constructed from many two-layered competitive networks (Rumelhart &

Zipser 1986). Furthermore, specialized paradigms such as the Kohonen feature map and counterpropagation include competitive layer substructures. Thus competitive learning is pattern-classification tool as well as a building block for larger, more powerful networks.

COMPETITION THROUGH INHIBITION

A competitive layer of processing units can be implemented using inhibitory connections, where each unit inhibits every other unit. The resulting architecture provides a way for the network itself to choose the unit that wins the competition. The activation levels of the processing units gradually "relax" to the point where the unit with the highest incoming sum remains activated. In contrast, the competitive learning model described above chooses the winning unit by using an algorithm that identifies the unit with the highest incoming sum.

Consider a single layer of units that corresponds to the second layer of a competitive learning scheme, shown in Figure 6-7. Suppose that each unit has already calculated its incoming sum, S_j. S_j becomes the initial activation value for unit j

$$A_j(0) = S_j = \text{activation level at time} = 0$$

Assume that each processing unit in this layer sends an inhibitory connection to every other unit in the layer. The activation level A_j of unit j is proportional to the strength of the inhibitory signal. A larger value of A_j thus results in a stronger inhibition being sent to all the other units. When a unit receives inhibition along any of these interconnections, its activation level A_j is decremented. Furthermore, each unit has an excitatory connection to itself, which increments its own activation level in proportion to its previous activation level.

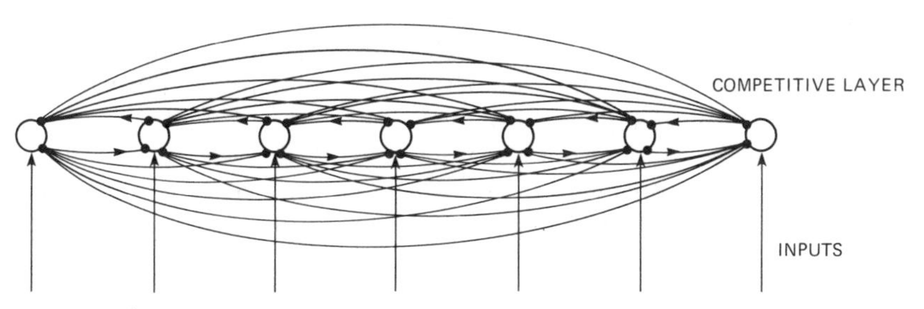

Figure 6-7. A competitive layer implemented with inhibitory connections.

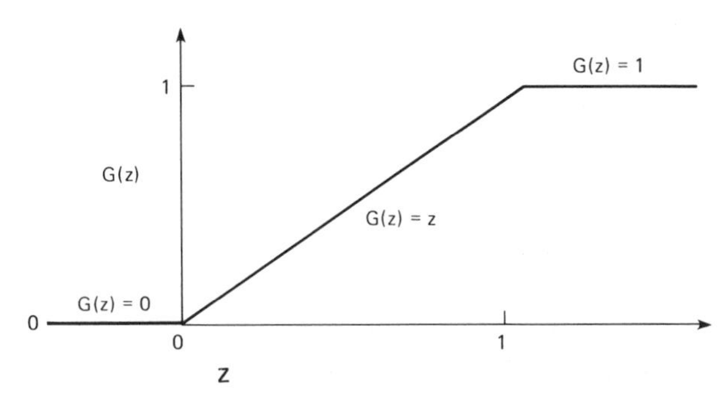

Figure 6-8. $G(z)$, a ramp function with hard limits at 0 and 1.

This process is implemented by

$$A_j(t) = G\left(A_j(t-1) + \frac{c}{M}A_j(t-1) - \frac{d}{M}\left[\sum_{\substack{i \\ i \neq j}} A_i(t-1)\right]\right) \quad (6\text{-}6)$$

where t = time step (e.g., 0, 1, 2, 3, . . .); $A_j(t)$ = the activation level for unit j at time t; c,d = small constants ($\ll 1.0$); M = sum of activation levels of all units in the competitive layer; and G = a transfer function with unit gain and hard limiting at 0 and 1. In this model, time moves forward in steps. At each step, the parameter value A_j for each processing unit is updated according to (6-6). Each time step represents an incremental step forward in time, and thus Eq. (6-6) models an active, continuous connection.

A further characteristic of this model is that activation levels are restricted to the range 0 to 1. Activation levels that would otherwise go out of this range are fixed at 0 or 1. Thus the function G in equation (6-6) has the form shown in Figure 6-8.

The first term in parentheses in (6-6) is the previous value of unit j's activation level. The second term represents excitatory feedback from unit j to itself. Note that in this term, the unit that was initially the highest gets the greatest increase due to this feedback. The third term sums inhibition from all other units, and subtracts a fraction of this sum from unit j's activation level. Thus the unit with the highest initial value is decreased the least.

Figure 6-9 gives results from a sample network that was subjected to the inhibitory relaxation specified in (6-6). The unit that had the highest value initially slowly reaches the point where it is the only unit with an activity level of 1. In the final state, the activity values of the remaining units are negligible. Parameters must be tuned appropriately to get the states of the winner to move to 1.0 and the losers to move to 0.0 effectively.

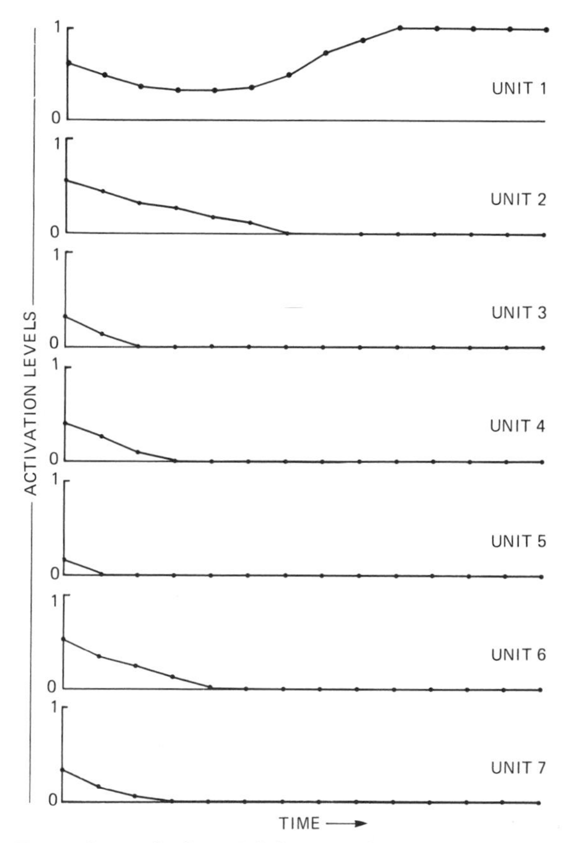

Figure 6-9. Example results from inhibitory relaxation of a competitive layer.

Implementation of the competitive layer as a layer of inhibitory units has certain advantages. The interconnections are updated in incremental steps, modeling a continuously varying connection. Such a connection is closer to biological reality than the competitive learning architecture described earlier in this chapter, as inhibitory connections are common in biological neural systems. Results from inhibitory competition models thus contribute to our understanding of biological systems. A further consideration is in hardware implementation. Continuously varying connections are required in some specialized designs for hardware based on analog circuits. The "winner take all" might be implemented in specialized hardware as an inhibitory layer, in which the layer "relaxes" to allow the winner to be identified. Such a hardware design would be limited by the density of interconnections required, however.

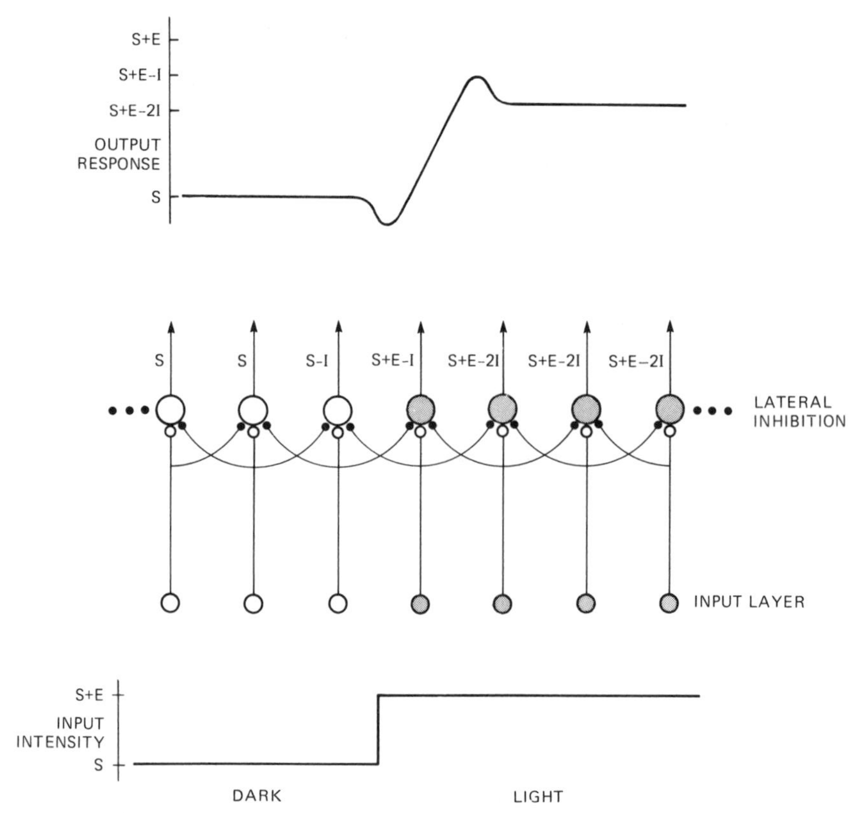

Figure 6-10. Model of a simple lateral inhibition network. Input intensity goes from dark to light. (Adapted from Bullock, Orkland, and Grinnell, ***Introduction to Nervous Systems.*** W. H. Freeman, Inc. 1977).

LATERAL INHIBITION

Lateral inhibition is a network architecture that is related to competitive learning, but whereas a competitive layer has units that inhibit all other units, a lateral inhibition layer has processing units that inhibit only those units that are nearby. Lateral inhibition appears in biology, in the visual and auditory systems, providing an ability to enhance contrast, peaks, and edges of incoming patterns.

Figure 6-10 shows a model that performs lateral inhibition, fashioned from systems observed in biology (Bullock, Orkand, and Grinnell 1977). In this model, each unit inhibits only its nearest neighbors. The input layer is a single row of units, and the stimulus is an edge that changes from light to dark. Each unit receives an amount of excitation S from the dark portion. The light

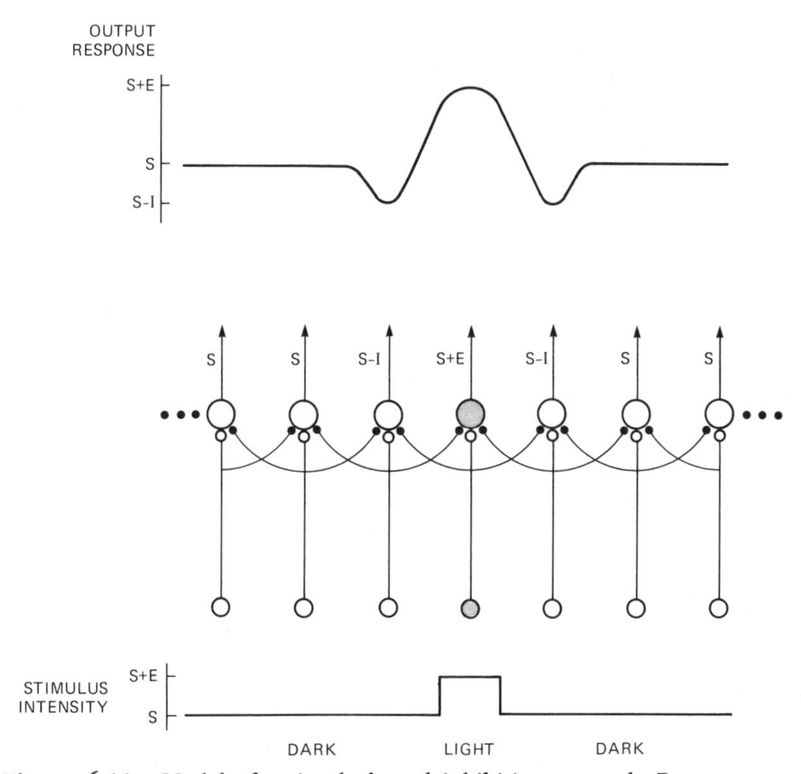

Figure 6-11. Model of a simple lateral inhibition network. Response to a stimulus consisting of a light region surrounded by darkened areas. (Adapted from Bullock, Orkland, and Grinnell, *Introduction to Nervous Systems.* W. H. Freeman Inc. 1977.)

portion gives an excitation of $S + E$. The amount of inhibition to neighbors is fixed at the value I.

The edge-enhancement properties of this network are apparent from its output, shown at the top. The borderline area of the light portion becomes lighter, and the borderline area of the dark portion becomes darker. The transition from dark to light appears enhanced. If the same network had its lateral inhibition connections removed, a step function would be the output. Note that the scale of the output pattern is independent of the scale of the input pattern and thus the actual magnitude of the change in the output cannot be compared to that in the input.

Figure 6-11 shows a similar network, this time with an input that consists of a light area surrounded by darker areas. The border before and after the light area is darkened further in the output; the light area itself appears lightened in the output. The response of the network is again to enhance transitions at the edges.

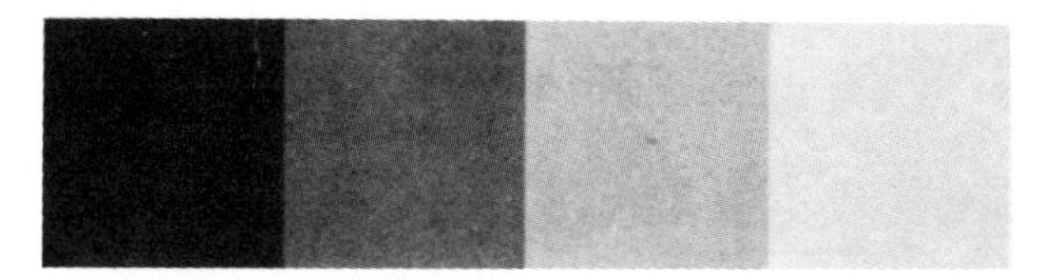

Figure 6-12. Mach bands (from Dowling, *The Retina.* Harvard Univ. Press, 1987).

Lateral inhibition appears in many biological systems, for example in the retina, where amacrine and horizontal cells provide inhibitory interactions to cells in close lateral proximity to one another. The auditory system also appears to make use of lateral inhibition for discerning and encoding pitches. Lateral inhibition appears responsible for the optical illusion known as Mach bands, shown in Figure 6-12. Adjacent squares have different shades of darkness. The area to the left of each border appears darker than the area farther to the left; the area to the right of each border appears lighter than the area farther to the right. This is due to contrast enhancement, a characteristic of lateral inhibition networks. If the borderline is covered with a narrow strip of paper, the illusion is eliminated.

Lateral inhibition can be used computationally to preprocess noisy data in order to emphasize peaks and increase edge contrast. Figure 6-13 shows

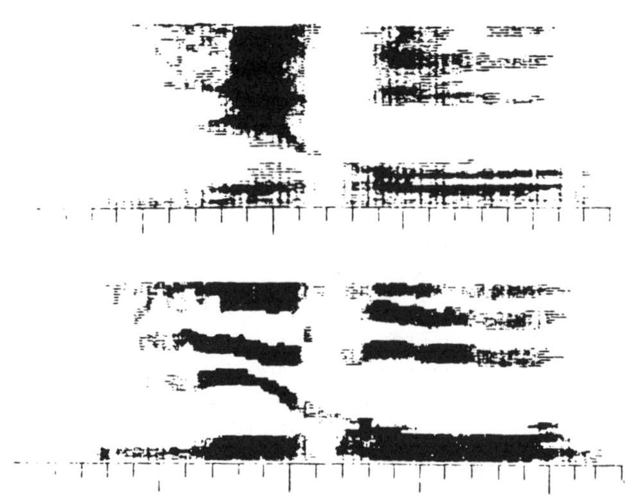

Figure 6-13. Action of lateral inhibition on a speech formant for the word "zero." (a) The original formant (frequency versus time). (b) The same data after subjected to a lateral inhibition process. (From Beroule, *ICNN Proceedings* © 1987 IEEE.)

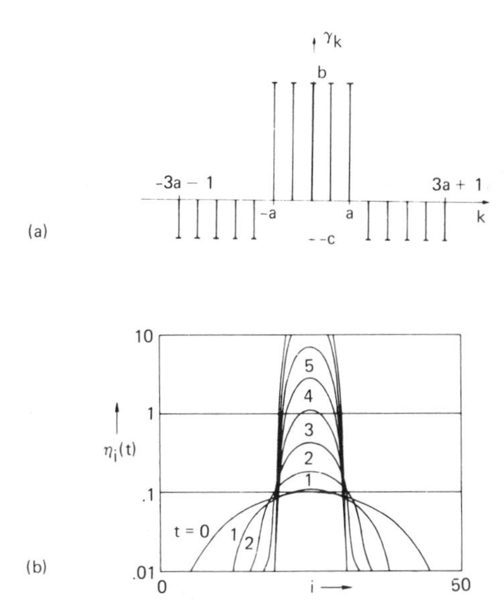

Figure 6-14. Lateral inhibition with an excitatory center. (a) Effect of the processing unit at $x = 0$ on nearby units. (b) Activation levels of units along the x-axis for successive time steps $t = 0, 1, \ldots, 5$. (From Kohonen, *Self Organization and Associative Memory*. Springer-Verlag. 1988.)

speech formants generated by the utterance of the word "zero." The x axis plots time, the y axis plots frequency, and the darkness reflects intensity. The top plot is from the original sound recording; the bottom shows the results after lateral inhibition is applied. The peaks are retained, with sharpened edges and improved focus.

Figure 6-14 shows a variation of lateral inhibition (Kohonen 1988). Each processing unit excites other units in close proximity. Units that are farther away are inhibited. The region of influence is eventually cut off. Figure 6-14b shows the time course of a simulation using this model. The units are placed along the x axis and their activation levels are graphed. Time 0 starts with a flat, broad peak. After five time steps, this peak becomes more focused, with a high and narrow profile.

Note that the competitive relaxation from (6-6) had each unit excite itself; the lateral inhibition model here has each unit excite itself and other units nearby. Use of the lateral inhibition network requires experimentation and tuning of parameters for the size and intensity of the excitatory area, and the size and intensity of the inhibitory areas.

Peaks in two dimensions can be similarly emphasized with a layer of units arranged on a plane. Each unit excites other units nearby, and inhibits units

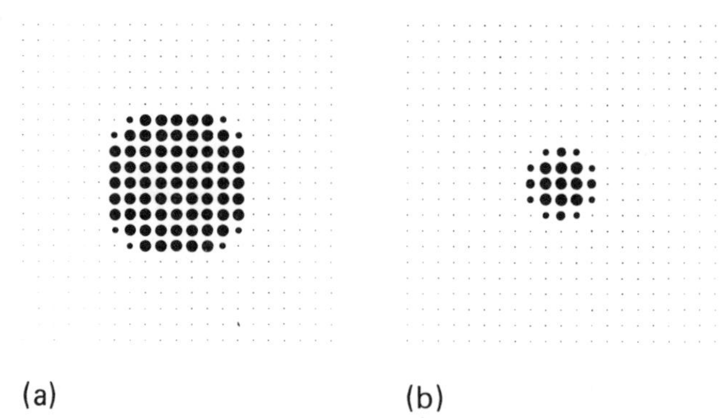

Figure 6-15. Clustering of activity in a two-dimensional array. (a) Positive feedback stronger. (b) Negative feedback stronger. (From Kohonen, *Self Organization and Associative Memory.* Springer Verlag. 1988.)

that are farther away. The strength of excitation and inhibition must be tuned. If excitation is strong, then an area of activation becomes large. If inhibition is strong, then an area of activation becomes small. Figure 6-15 shows examples of these two cases, taken from a starting point where a cluster of activity exists at the center of the grid. When positive feedback is stronger, the results depicted in Figure 6-15a occur after a period of time. When negative feedback is stronger the outcome shown in Figure 6-15b results.

References

Beroule, D. 1987. Guided propagation inside a topographic network. *IEEE ICNN 1987* IV-469–IV-476.

Bullock, T. H., R. Orkand, and A. Grinnell. 1977. *Introduction to Nervous Systems,* San Francisco: W. H. Freeman.

Carpenter, G. A., and S. Grossberg. 1988. The ART of adaptive pattern recognition by a self-organizing neural network. *Computer* (March): 77–88.

Dowling, J. E. 1987. *The Retina.* Cambridge, Massachusetts: Harvard University Press.

Grossberg, S. 1988. *Neural Networks and Natural Intelligence.* Cambridge, Massachusetts: MIT Press.

Kohonen, T. 1988. *Self-Organization and Associative Memory.* New York: Springer-Verlag.

Rumelhart, D. E. and D. Zipser. 1986. Feature discovery by competitive learning. In *Parallel Distributed Processing,* eds. D. E. Rumelhart and J. L. McClelland and PDP Group. pp. 151–193. Cambridge, Mass.: MIT Press.

Stork, D. G. 1989. Self-organization, pattern recognition, and adaptive resonance networks. *J. Neural Network Computing.* 1(1):26–42.

7

The Brain and Its Neurons

The fascination of neural network computing arises in part from its relationship to the brain. Claims have been made in recent years that artificial neural networks explain the basic mechanisms and dynamics of the brain. In fact, although the name neural networks is borrowed from the term for the brain's networks of biological nerve cells, a key point in understanding artificial neural networks is that present models depart dramatically from biological structure and cognitive processes. Vast differences exist between these two types of systems, both in architecture and capabilities.

The human brain has billions of individual nerve cells (neurons) and trillions of interconnections. Neurons are continuously processing and sending information to one another. The structure of these cells and the anatomy of their interconnections are extremely complex compared to current artificial neurocomputing models. Furthermore, the human brain has amazing capabilities: the ability to recognize patterns and relationships, store and use knowledge, reason and plan, learn from experience, and understand what is observed. No models to date have been able to duplicate the incredible powers and profound capacities of the human brain.

Furthermore, the brain appears to harbor that elusive quality called "the mind." No artificial model has been able to predict any significant part of the personal or emotional experience of the human mind. Our models do not explain how the known anatomy and physiology of the brain generates our cognitive and perceptual experience.

A description of neural network architectures would not be complete, however, without including biological neural systems. Similarities between these two types of systems are important to observe, both to learn more about the nervous system and to gather ideas for neurocomputing architectures.

Both systems are based on parallel computing units that are heavily interconnected and both systems include feature detectors, redundancy, massive parallelism, and modulation of connections.

However the differences between biological systems and artificial neural networks are substantial. Artificial neural networks usually have regular interconnection topologies, based on a fully connected, layered organization. While biological interconnections do not precisely fit the fully connected, layered organization model, they nevertheless have a defined structure at the systems level, including specific areas that aggregate synapses and fibers, and a variety of other interconnections. Although many connections in the brain may seem random or statistical, it is likely that considerable precision exists at the cellular and ensemble levels as well as the system level. Another difference between artificial and biological systems arises from the fact that the brain organizes itself dynamically during a developmental period, and can permanently fix its wiring based on experiences during certain critical periods of development (Hubel and Wiesel 1979). This influence on connection topology does not occur in current artificial neural networks.

Dynamic processing is also different. Artificial neural networks usually exhibit updates that are simultaneous or at regular intervals. Biological systems are continually adjusting the characteristics of each neuron, and they perform sudden, asynchronous updates when initiating or receiving impulse signals. Processing of incoming signals by artificial neural units is relatively simple, usually involving a summation and thresholding of scalar inputs. Processing at synapses in biological systems is much more complex, and involves a detailed microstructure with participation of many different biochemical components.

The future of neurocomputing can benefit greatly from biological studies. Structures found in biological systems can inspire new design architectures for artificial neural models. (These new architectures must be researched to determine what computational advantages they might—or might not—offer.) Similarly, biology and cognitive science can benefit from the development of neurocomputing models. Artificial neural networks do, for example, illustrate ways of modeling many characteristics that appear in biology, in the human brain. Conclusions, however, must be carefully drawn to avoid confusion between the two types of systems.

BRAIN ORGANIZATION

The human brain contains about a hundred billion (10^{11}) nerve cells, each with as many as 10,000 interconnections with other nerve cells. These neurons are densely packed in a highly interwoven mass of tissue that looks, upon dissection, like a web of fibers. The sheer numbers of neurons and the high density of interconnections account for a substantial part of the brain's computational power.

Processing in the brain is probably broken down into groups or assemblies of neurons that act as functional units. Such an assembly of neurons might be active simultaneously or might create particular patterns of activity as a group. The number of such assemblies, about $2^{(10^{11})}$, is an astounding amount of possibilities to have available, each with potential processing significance. In fact, the number of possible assemblies is greater than the number of atomic particles that make up the known universe. (Current estimates are 10^{100} particles).

The human brain, diagrammed in Figure 7-1 from the side, is covered by the cerebral cortex, the convoluted outer layer. The size and complexity of the cerebral cortex constitutes a critical difference between humans and lower animals. In the past three million years of evolution, an explosion in brain size has taken place, primarily from an expansion in the size of the cerebral cortex. The structures below the cortex were retained with fewer changes.

Different regions of the cortex are specialized for complex tasks such as speaking, understanding speech, analyzing visual information, organizing motor activity, and other aspects of intelligent behavior. For example, in right-handed people, damage to a part of the left side of the cerebral cortex can destroy the person's ability to speak correctly or to understand speech.

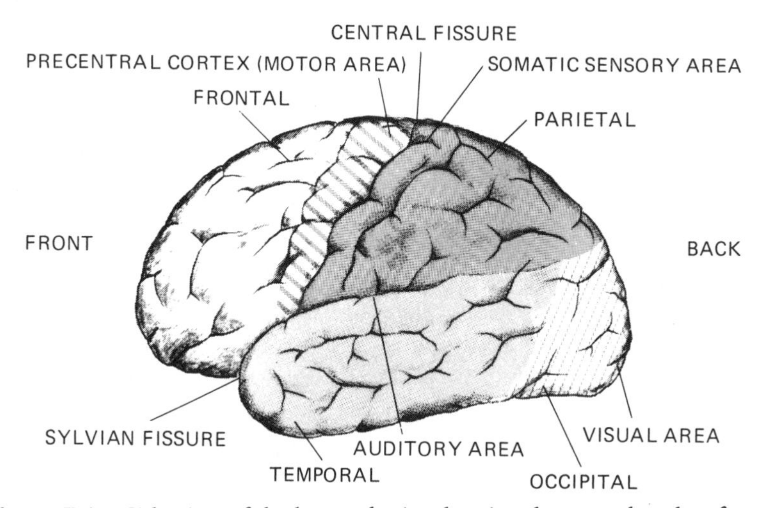

Figure 7-1. Side view of the human brain, showing the convoluted surface of the cerebral cortex (From *The Brain: An Introduction to Neuroscience.* By Richard F. Thompson. © 1985 by W. H. Freeman and Co. Reprinted with permission.)

Damage to the right side can cause difficulty in performing spatial tasks, such as navigating or understanding complex diagrams.

Sensory and motor systems have pathways to specific areas of the cerebral cortex: Visual information traveling to the occipital lobe in the back (see Figure 7-1), information from the skin and body to a middle region (the parietal lobe), and auditory information to the auditory cortex, located in the upper part of the temporal lobe. In front of the central fissure is the motor cortex, which participates in the control of muscular movements.

Each sensory area of the cerebral cortex contains a map that reflects the layout of the corresponding sensory organs. The eye, for example, projects impulses to a map of the surface of the retina in the back of the cortex. The map of the ear on the cortex reflects the sheet of receptors in the cochlea, the part of the inner ear that receives auditory input. Different receptor cells in the cochlea respond to different frequencies of sound. Thus, there is a map of sound frequencies along the surface of the auditory cortex. In a similar fashion, sensory areas along the body are laid out on the surface of the somatic sensory area.

The association cortex, a large part of the cerebral cortex, appears to be involved in higher brain function, although the specific functions and mechanisms are not fully understood. The size of the association cortex is considerably larger proportionally in humans than in lower animals.

Figure 1-3 illustrates the major neural structures that are placed within or below the cerebral cortex. These structures include the cerebellum, the spinal cord, the brain stem, the thalamus, and the limbic system. The cerebellum is involved with sensory-motor coordination, including precise muscular activities such as running, jumping, and playing a musical instrument (Albus 1981). The spinal cord, below the brain, functions in reflexes and passes signals up and down between the body and the brain. The brain stem, above the spinal cord, is concerned with respiration, heart rhythm, and gastrointestinal function. The thalamus — densely packed with synapses — is located near the center of the lobes, and functions as a relay station for the major sensory systems that project to the cerebral cortex.

The limbic system, which includes the amygdala, the hippocampus, and adjacent regions, is an evolutionary continuation of the primitive brains of early vertebrates such as reptiles. Much of the limbic cortex is involved with olfaction (smell), and participates in finding appropriate responses to various smells in the environment. In higher animals, the hippocampus may have taken on new roles, such as involvement in learning and memory.

The brain is densely packed, and its subsystems fit together like three-dimensional puzzle pieces that leave no space empty between them. Structures such as the amygdala, the thalamus, and the hippocampus curve about in unusual shapes, and fibers weave their way around, sometimes forming

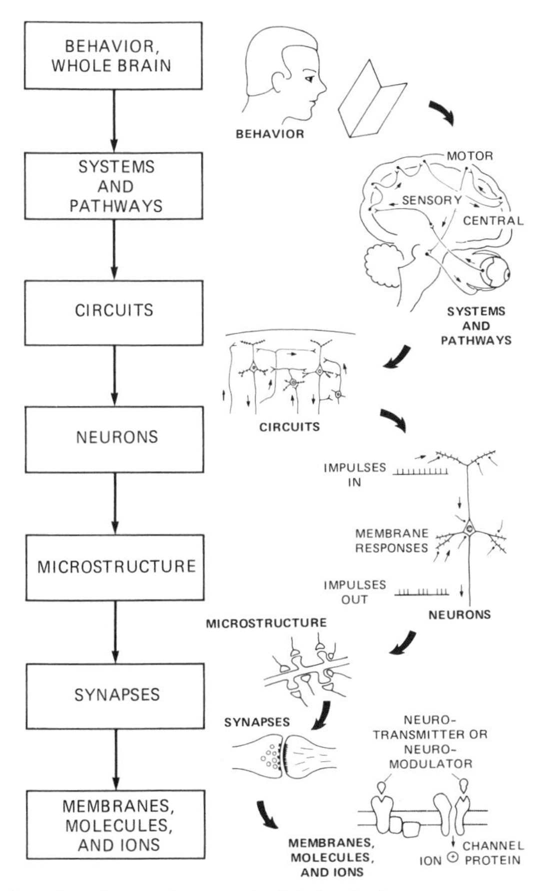

Figure 7-2. Levels of organization in biological nervous systems (adapted from Shepherd. *Neurobiology, 2nd ed.* Oxford Univ. Press 1988).

sheaths of parallel strands and sometimes reaching globular areas dense with synapses and cell bodies. Somehow within this complex structure exists a vast network of subunits and interconnections that are acting continuously, and are responsible for emergent intelligent properties.

Figure 7-2 illustrates the many levels of organization in biological nervous systems. At the top is behavior—the ongoing activities of an individual. Behavior is determined at the "whole brain" level. The second level consists of systems and pathways. The visual system, the auditory system, and the motor system taken as a whole each fit into this category. The third level is local circuits—smaller circuits in the brain that are responsible for local processing. These are assemblies of interconnected neurons. The next level is

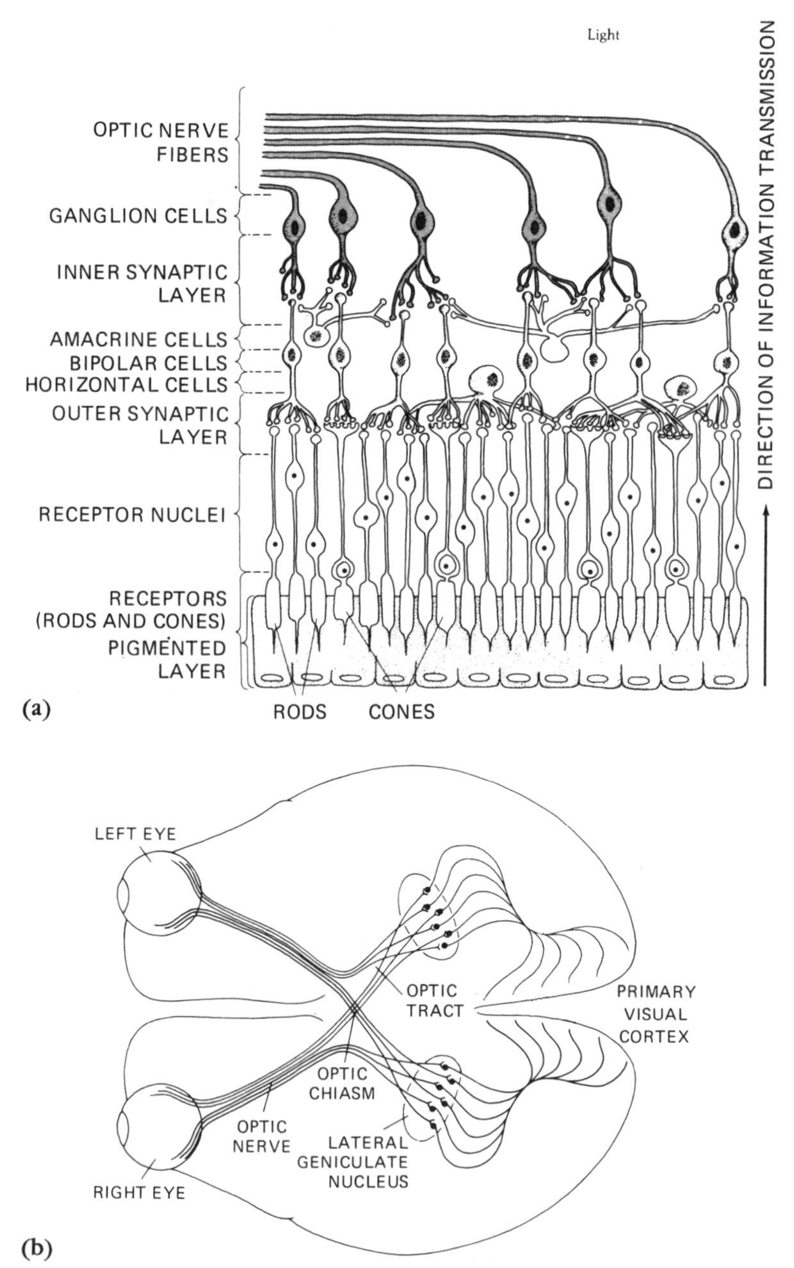

Figure 7-3. (a) A cross-section of the retina showing different layers of cells. Light impinges from the top, is received by rod and cone cells at the bottom which initiate signals that eventually stimulate the ganglion cells (top). (From Cornsweet. *Visual Perception.* 1970. Academic Press.) (b) Basic pathways of the visual system, from the eye to the primary visual cortex (from Hubel and Wiesel. Brain mechanisms of vision, *Scientific American* Sept. 1979).

the individual neuron. Each neuron has an input area, a cell body, and an output area that sends signals to other neurons. Below the level of the neuron are microcircuits — structures that affect the areas around synapses. The next smaller level consists of synapses and junctions — places where cells transmit signals from one to another. Below the synaptic level is the participation of membranes, molecules, and ions in the transmission of signals.

The visual and auditory systems each have a sensory organ that acts as a transducer to connect stimulus energy (photons, sound waves) into neural signals for subsequent processing in the brain. There are one or more nuclei (clusters of neurons), where the incoming fibers interconnect to new nerve cells, which in turn carry their signals farther into the brain. A topological map appears at the cerebral cortex level.

Figure 7-3a shows the retina, the visual system's sensory organ. The retina is composed of five layers of neurons. First, receptor cells respond to light signals from outside. Bipolar cells interconnect to receptors, carrying their signals to the ganglion cells, which transmit signals to the brain. At each interconnection stage some horizontal processing occurs. The transition from receptor cells to bipolar cells is influenced by horizontal cells, which provide lateral interactions. Farther back, amacrine cells also provide lateral interactions between the different cells that transmit signals back toward the brain. As a result, the signals from the ganglion cells to the brain encode local areas of the visual signal, and thus become a type of feature detector.

The optic nerve carries the ganglion cell signals back to the brain, where they branch and interconnect in a structure called the lateral geniculate (see Figure 7-3b). New neurons that originate at the lateral geniculate carry signals farther up to the visual cortex, at the back surface of the brain. It is here that the brain makes a "map" of the visual field (such topological maps are discussed later in this chapter).

The major structures of the auditory system are shown in Figure 7-4. Receptors are located in the cochlea, a coiled structure filled with fluid (shown in Figure 7-4a). If the coil were unwound, the basilar membrane inside — ennervated by receptor cells along its length — would flatten out. An incoming sound sets up a pattern of waves on the basilar membrane so that some regions distend more than others. These regions then stimulate their receptor cells, which in turn activate fibers of the auditory nerve that go to the brain.

A simplified sketch of the auditory pathways (Figure 4-7b), starting where the auditory nerve enters the brain, shows where the signals branch and interconnect at four major brain structures: the cochlear nucleus, the superior olive, the inferior colliculus, and the medial geniculate. Each of these structures is rich in interconnections and serves to relay and process information from the ears. After the signal reaches the first station — the cochlear nucleus — some outgoing signals cross to the opposite side of the brain, so

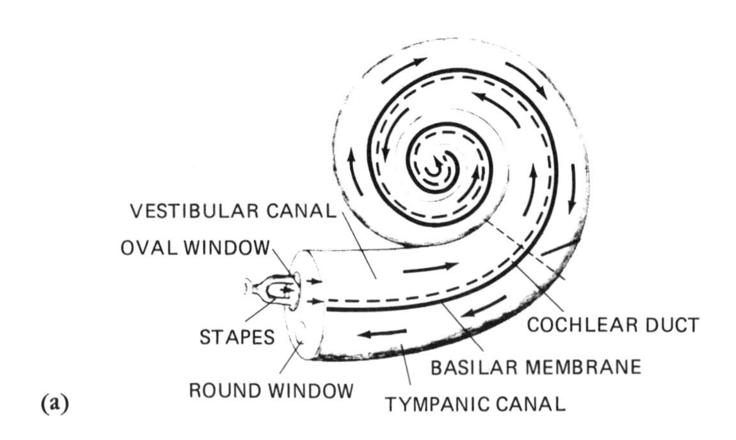

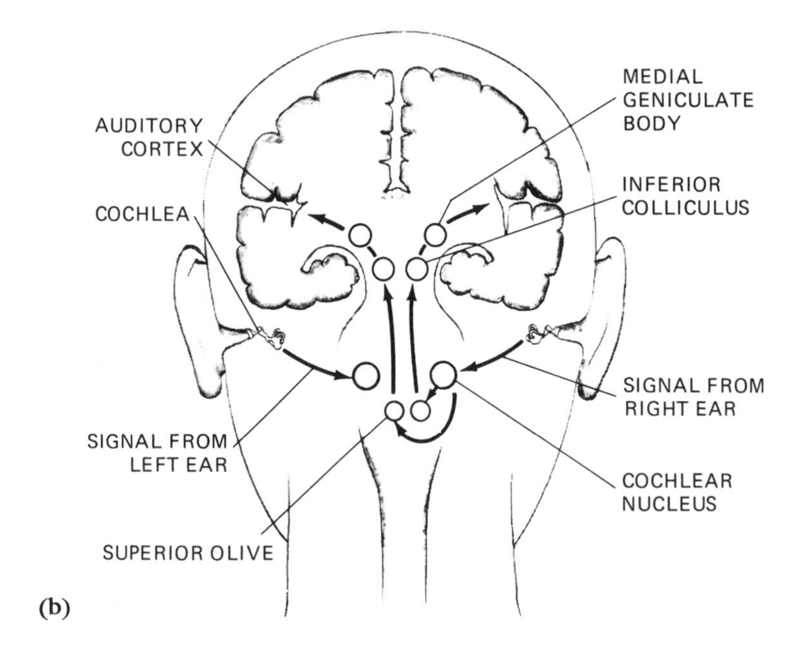

Figure 7-4. (a) The cochlea, a coiled structure in the inner ear that generates nerve signals to the brain in response to incoming sounds. The cochlea contains the basilar membrane, which forms waves that stimulate the auditory neurons. (b) Basic pathways of the auditory system, from the ear to the auditory cortex. (From P. Lindsay and D. Norman. *Human Information Processing: An Introduction to Psychology, 2nd Ed.* © 1977 Harcourt Brace Jovanovich.)

that upper-level processing can be *bilateral,* with input from both ears. From the highest substructure—the medial geniculate—fibers project to the auditory region of the cerebral cortex.

An important feature of sensory and motor systems is the existence of topological maps in the brain. For example, there are multiple representations or maps in the cortex that correspond to particular parts of the body for somatosensory or motor activities, or to a part of the extrapersonal sensory world. In the case of the visual system, much of the visual area of the cortex is ordered in such a way that stimuli from nearby positions on the retina activate nearby positions on the surface of the cortex. In the auditory system, areas of the auditory cortex are organized so that similar tones elicit activity in nearby or juxtaposed positions. In the motor map, nearby positions activate muscular areas that are close to one another. A continuum is somehow built into this ordering so that movement across the surface of a map corresponds to a logical progression of stimuli or motor activities.

Topological maps tend to dedicate greater area to activity that is utilized more. For example, the fovea of the retina is the central area of the retina and is more highly ennervated; visual acuity is best there. The retinal map on the cortex dedicates more area to the fovea.

Figure 7-5a shows the map of the retina on the visual cortex at the very back of the brain. Half of the visual field is charted on the right, with its different areas coded to match the corresponding areas of the map on the left. The general shape of the back tip of the brain and its fissures and convolutions are apparent from the drawing. The area of cortex dedicated to the fovea appears at the very back tip of the cortex, and is disproportionately large compared to the other areas. One can also observe the elastic distortion that must be made of the visual field in order to fit it onto the irregular bulges and convolutions of the cortex.

To illustrate a topological map of the auditory system, Figure 7-5b shows the tonotopic map of the mustache bat. For this creature, the frequency of 61 kHz has great importance because the bat does echolocation by emitting a pure tone of 61 kHz and analyzing the return. Its auditory system can discern subtle differences around this frequency to infer locations of objects from the returning echoes. The tonotopic map has a large central region dedicated to 61–62 kHz sounds, and side regions dedicated to lower and higher pitches. In contrast, the cat, which does not use echolocation, has a more regular pattern of activity that spans 0–32 kHz, but is also continuous along increasing pitch lines. (See Figure 7-5c.)

Additional, more complex structures are superimposed on these topological maps. For example, the visual system is also organized by cortical columns, ones that are perpendicular to the surface of the cortex and are each dedicated to particular types of stimuli. A cortical column may be dedicated

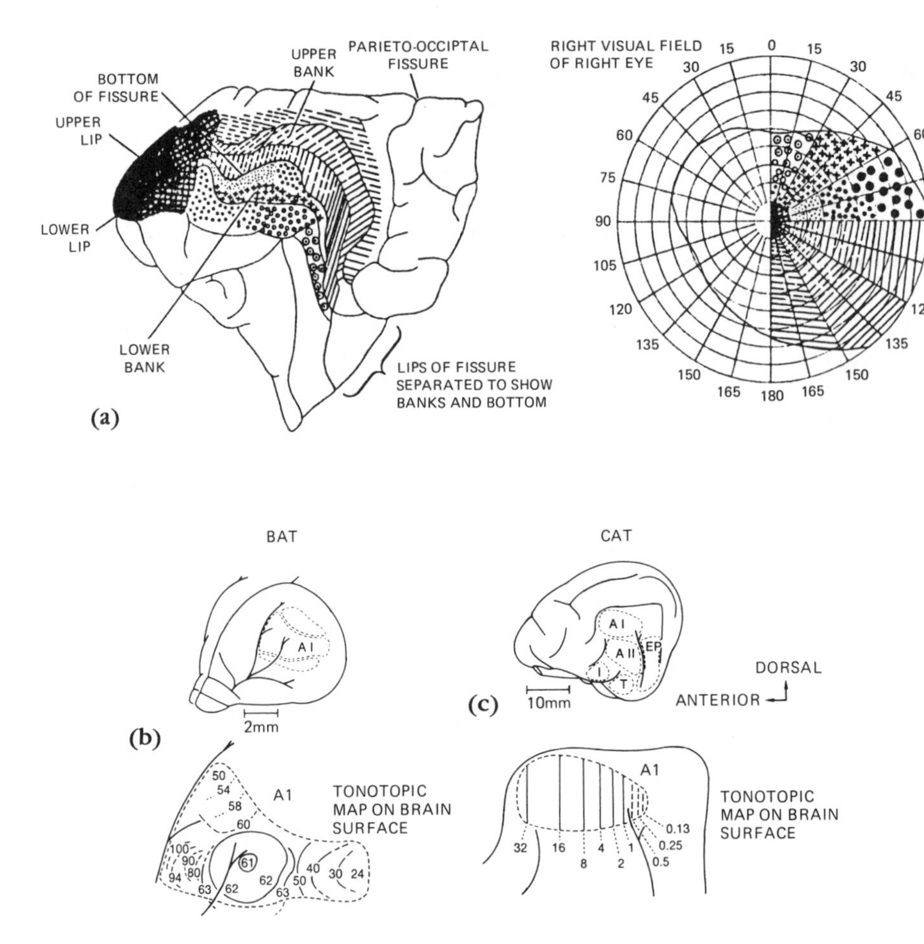

Figure 7-5. (a) Left: Occipital lobe of the brain, which consists of the back tip of the cerebral cortex (the visual cortex), side view. Right: the visual field, divided into different areas. Areas with like shading in the two pictures correspond; a visual stimulus in the visual field in each shaded area activates a place on the occipital lobe shown with like shading. (From M. Glickstein. The discovery of the visual cortex. *Scientific American* Sept. 1988.) (b) TOP: Side view of the bat brain, showing regions of the auditory cortex. BOTTOM: A tonotopic map of a region of the bat auditory cortex. (c) TOP: Side view of the cat brain, with regions of the auditory cortex. BOTTOM: The tonotopic map found in the A1 region of the auditory cortex of the cat. (From: Suga. Specialization of the auditory system for reception and processing of species-specific sounds. *Fed. Proc.* 1978; based on data from Woolsey, Merzenich and Suga.)

to a particular eye (an ocular dominance column); this type of column receives stimulus from only one eye. A smaller column could be dedicated to a particular orientation of an edge in the visual field; columns of this type respond to different angles for edges or bars of light.

NERVE CELLS—BASIC STRUCTURE

The schematic diagram in Figure 7-6 shows the elemental building block of the nervous system—an individual nerve cell, or neuron. A typical nerve cell has three major regions: the cell body, the axon, and the dendrites. The cell

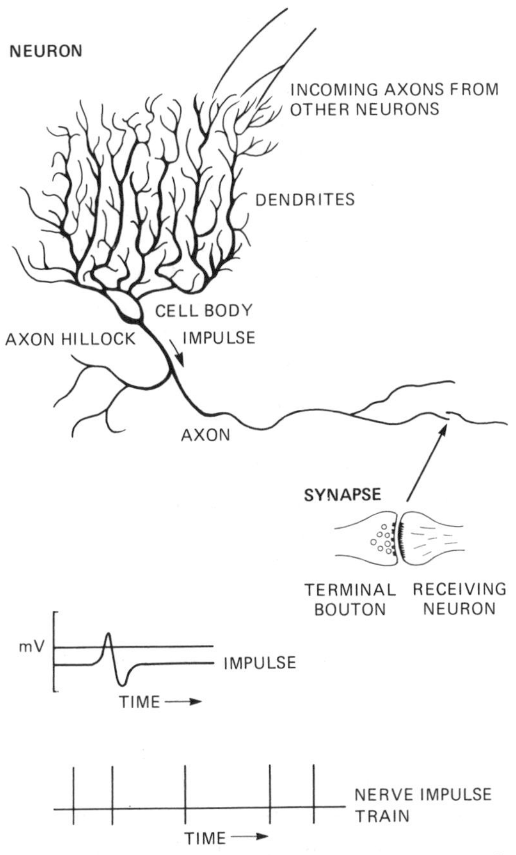

Figure 7-6. Recording a nerve impulse train with a microelectrode placed near an axon. A trace of the nerve impulse waveform is shown, as it would appear on an oscilloscope if recorded from the microelectrode. A nerve impulse train is shown at bottom.

body, or soma, provides the support functions and structure of the cell. The axon is a branching fiber that carries signals away from the neuron, and the dendrites consist of more branching fibers that receive signals from other nerve cells.

The axon, as the output mechanism for the neuron, conducts signals away from the cell to other cells via interconnection points called synapses. Although there is only one axon for each cell, it can branch tremendously and thereby send separate branches to different locations. Thousands of branches and interconnection points are possible for a single axon. The outgoing signals are pulses, which are usually initiated at the axon hillock, where the axon connects to the cell body.

Terminal boutons—shown magnified in Figure 7-6—are at the ends of the axon branches. These structures, which usually form small bulges at the end of each fiber, release chemical transmitters, which cross the synaptic gap to reach a receiving neuron. When these transmitters are released, the membrane of the target neuron is affected, and its inclination to fire its own impulse is changed—either increased or decreased, allowing an incoming signal to be either excitatory or inhibitory.

It is possible to record electrical pulses going down a single nerve axon by placing a finely ground microelectrode near the neuron in a living brain. Although each pulse from that neuron has the same waveform and velocity, each is initiated at a unique time. Thus the times that nerve impulses occur are key data to record. Timing in different impulse trains may be very different, and can be analyzed statistically. Figure 7-6 shows an example impulse waveform and a hypothetical nerve impulse train.

Dendrites are branching processes designed to receive incoming signals from other nerve axons via synapses. Dendrites can be densely branched, and can grow from one or more different locations of a cell body. Typically these processes taper, with smaller diameters near the terminal ends. The sides have dendritic spines, which are protruding structures important for effective reception of incoming signals. The diameters of the branches and trunks of dendrites are significant in determining how incoming signals are summed and processed. Although the most important function of the dendrites is to receive signals, dendrites of some neurons such as those in the olfactory bulb may transmit as well as receive signals via synapses.

The cell body or soma contains the nucleus (the carrier of genetic material) and apparatus for support of basic life processes of the neuron, such as energy generation and protein synthesis, and provides structural support to the cell. Cell bodies can act as information processors, as they sometimes receive input from other neurons via synaptic connections. The soma also exchanges nutrients and other compounds with the axon, as an active transport system exists to move nutrients and other compounds from the cell body down the axon and from the terminal end back to the cell body.

THE DIVERSITY OF NERVE CELLS

Nerve cells come in many different sizes and shapes, and are usually tailored for the jobs that they do: Some nerve cells have tremendous arborizations of dendrites or axons, some have long branches, and others have short branches. Neurons usually create signals via impulses, but some transmit signals to other cells without an impulse (Dowling 1987). Some nerve axons are insulated with a material called myelin, and others are not. Trunk diameters of dendrites can differ from cell to cell, and the detailed structure of the axon hillock can be different, leading to different dynamics for impulse initiation.

Figure 7-7 shows a Purkinje cell, known for its large dendritic tree and its massive number of interconnections. The dendritic tree of the Purkinje cell can receive upward of 200,000 incoming synapses. Its arborizations are relatively flat, and lie at right angles to a series of parallel fibers. Purkinje cells are located in the cerebellum, and participate in balance and motor coordination, sending their outputs deep into the cerebellum (Shepherd 1988; Kanerva 1988). The Purkinje cell body is the darkened region at the bottom of the figure, with its axon going off to the lower right. A few initial branches of the axon can be seen.

In contrast to the Purkinje cell, Figure 7-8 shows a cell from the cerebral cortex. The axons and dendrites of this cell are too short to project beyond a local area, and the dendritic tree has considerably fewer branches than the

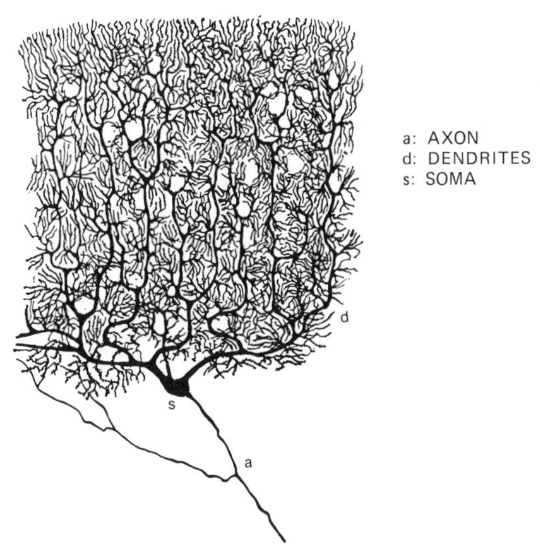

Figure 7-7. Purkinje Cell (from the cerebellum) with its dendrites stained; they form a flat fan. (From Cajal, 1909.)

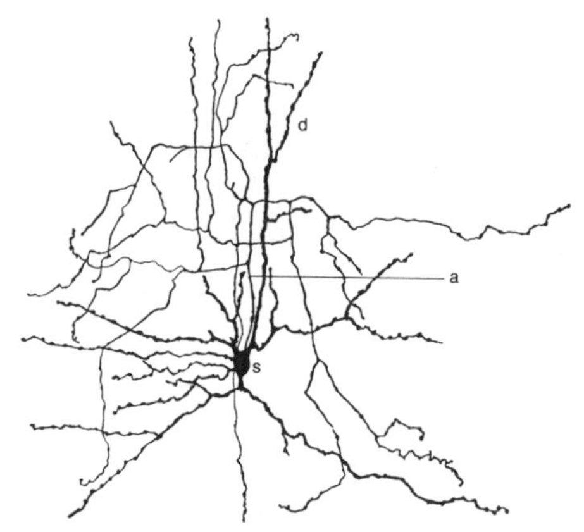

Figure 7-8. Cell from the cerebral cortex, with a short axon that does not project beyond the local region. (From Cajal, 1909.)

Purkinje cell. Cells like the one in Figure 7-8 provide communication between other local cells in the cortex.

Figure 7-9 shows a motor neuron taken from the electric ray. The neuron shown goes to the electric lobe of the fish. The axon and axon hillock are clearly shown, as well as the insulating myelin sheath (darkened area). Myelin insulation is particularly important when a nerve axon needs to conduct signals a long distance quickly, because myelin speeds the rate at which the impulse travels down the axon. Many animals have myelin insulation for

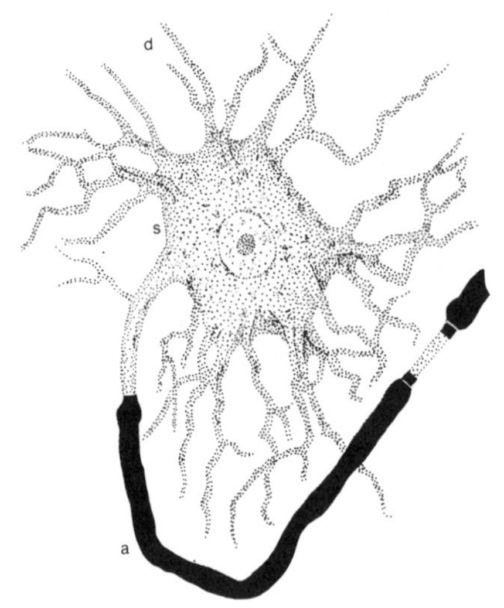

Figure 7-9. Motor neuron of the electric lobe of the electric ray. (From Cajal, 1909.)

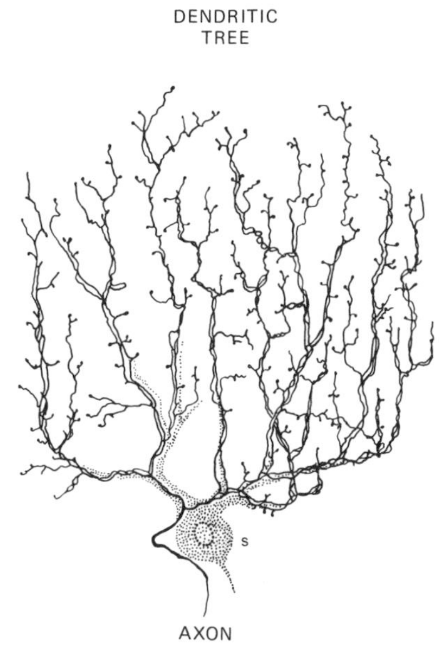

Figure 7-10. Purkinje cell, lightly shaded, with a climbing fiber—an axon from a distant cell body—entwining the dendrites. (From Cajal, 1909.)

motoneurons going from the brain to muscles in the extremities, for just that reason.

Some cell structures are organized to provide contact between pairs of cells. Figure 7-10 shows an axonal climbing fiber from a distant cell that has grown to match the dendritic tree of the Purkinje cell. Many synapses can then be formed between the climbing fiber axon and the target Purkinje cell. The basket cell, shown in Figure 7-11 has extended its axon past a line of somas of

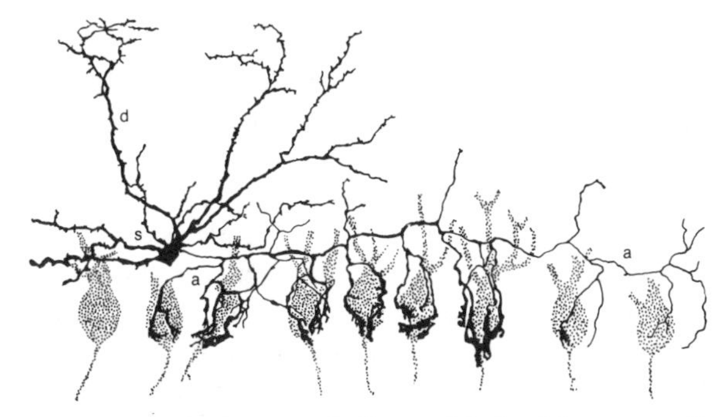

Figure 7-11. Basket cell of the cerebellum with dendrites extending upwards and an axon forming basket-shaped arborizations about a row of lightly shaded Purkinje cell bodies. (From Cajal, 1909.)

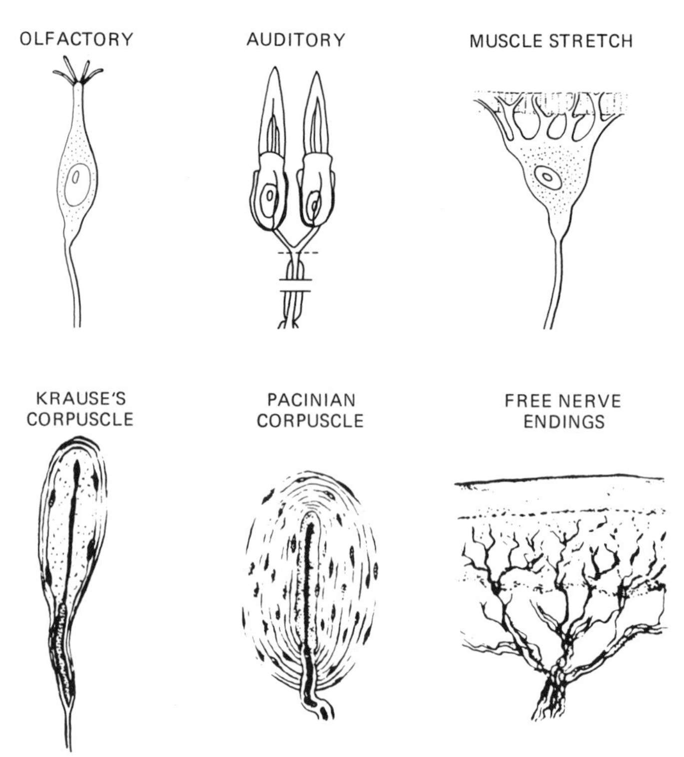

Figure 7-12. Receptor cells for smell, hearing, muscle movements, pressure and touch. (From Bodian. Neuron junctions. *Anatomical Record,* Vol 174. Alan R. Liss 1972; Guyton 1976; and Carlson.)

Purkinje cells. Axonal branches then grew to surround the cell bodies, so that they could intensify their interaction with the Purkinje cell by making many synapses to the cell body. The dendrites of the Purkinje cells are not shown in Figure 7-11, but the picture includes the sparse dendritic tree of the basket cell at the top. The climbing fiber and the basket cell illustrate some of the nervous system's mechanisms for creating strong connections with many synapses between cells.

Figure 7-12 shows a variety of receptor cells, including free nerve endings, a Pacinian corpuscle, hair cells from the vestibular system and auditory systems, and the muscle stretch receptor. In each case, environmental stimuli such as movements cause changes in the membrane of the receptor cell, which in turn cause the cell to initiate signals to the brain. The receptor cell sends an axon directly to the brain or it may interconnect to other neurons that eventually send an axon to the brain. Activity in different nerve fibers results in different sensations, for example, the brain knows that touch has occurred

Figure 7-13. Growth of the dendritic trees and axon branches of cortical pyramidal cells in the human, from fetus to adult. (Courtesy of Sidman and Rakic 1982, and Poliakov in Sarkisov and Preobrazenskaya 1959.)

because a particular nerve fiber connected to a touch receptor at that point is carrying a train of impulses. Specific sets of fibers are activated by noxious stimuli; others are sensitive to warmth and cold. Note that overlap can occur among the receptive fields (areas to which each receptor is sensitive). Firing rates and sometimes temporal structure of pulses are responsible for encoding signals up to the brain.

The previous examples showed a variety of different specialized neurons, with different branching structures, receptor mechanisms, and interconnection strategies. Next we show examples in which branching complexity increases.

Figure 7-13 shows neuronal growth from fetus to adult for pyramidal cells in the human cortex. Not only do dendritic and axonal branches grow longer throughout this development, but the number of branches increases. Figure 7-14 illustrates the increase in complexity over the evolutionary tree. Cells from the same locations are depicted for a variety of animals, starting with a primitive example—the lamprey—and progressing to more advanced animals, up to man. The number of branches and branch points on the dendritic tree increased with evolutionary development.

Other neurons are known for their feature-detecting abilities. A neuron structured to detect a temporal delay in incoming signals is depicted in Figure 7-15. Neurons with two distinct dendritic arborizations are found in the superior olive of the auditory pathway. The two dendritic trees extend from opposite sides of the cell soma. Note that the superior olive is actually two

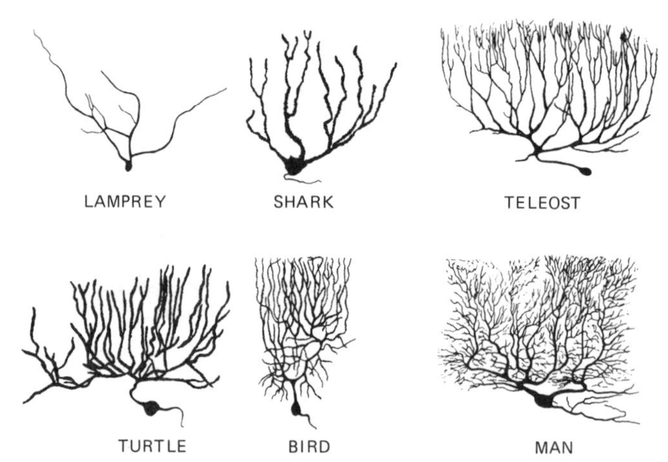

Figure 7-14. Purkinje Cells from a variety of different organisms. More primitive animals have a simpler branching structure. (From Nieuwenhuys 1969.)

structures, one on each side of the brain. In these neurons, one dendritic tree receives inhibitory input from the ear on the same side, and the other receives excitatory input from the ear on the opposite side. These cells appear to be involved in determining the origin of a sound source. When a sound is presented abruptly from the opposite side, the cells are excited (temporarily) because the signal arrives at the excitatory dendritic tree branch first. If a sound is presented from the same side, the cell is inhibited because the inhibitory connections are activated first and suppress the delayed excitatory signals. The cells can go from strong excitation to strong inhibition in a period of only a few hundred microseconds.

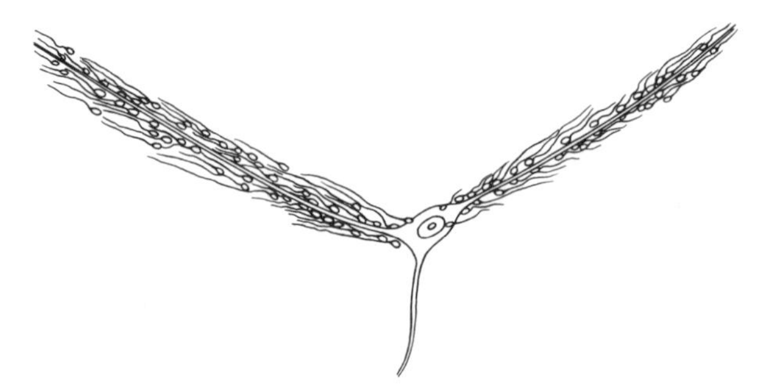

Figure 7-15. Cell from the superior olive of the auditory pathway, with two separate dendritic trees that serve to detect delays (in the onset of sounds) between the two ears. (From Stotler. An experimental study of the cells and connections of the superior olivary complex of the cat. *The Journal of Comparative Neurology*. Alan R. Liss. 1953.)

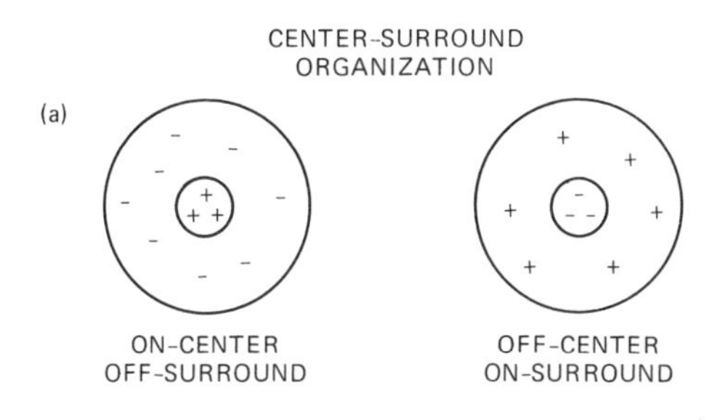

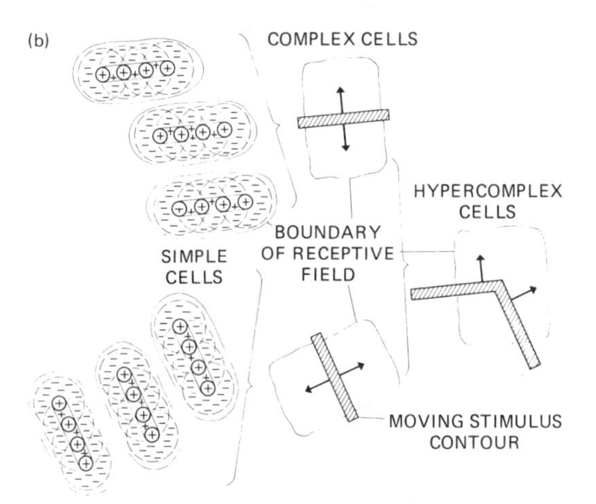

Figure 7-16. (a) Center-surround organization of the receptive field of simple cells in the visual pathway. These responses are typical of retinal ganglion cells and of cells in the lateral geniculate. (b) Receptive fields and stimulating patterns for complex cells in the visual pathway that process information from simple cells with center-surround organization. (From *Introduction to Nervous Systems,* by T. H. Bullock, et al. © 1977 by W. H. Freeman and Co. Reprinted with permission.)

Some of the most striking examples of feature-detecting neurons in biological nerve networks come from the visual system. Starting with the ganglion cell, which sends visual input from the retina to the brain, different cells in the visual system tend to respond to identifiable features in the visual field. Figure 7-16a shows the classic center-surround organization found in ganglion cells and some cells farther up in the brain. The total receptive field is the area of the visual field to which the cell responds. In the "on" center, a

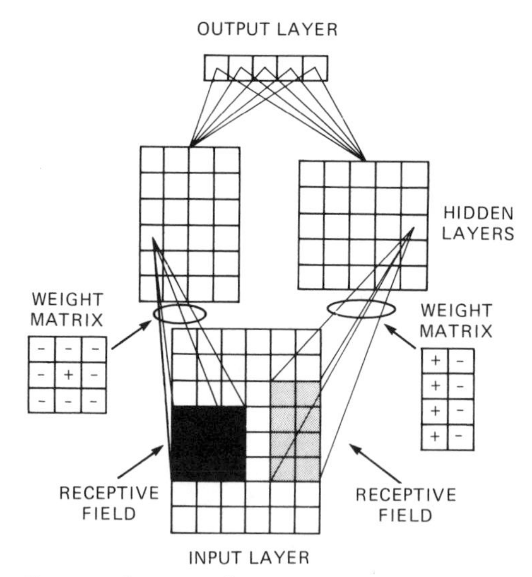

Figure 7-17. Feature detectors from an artificial neural network, similar in shape to feature detectors along biological visual pathways shown in Figure 7-16. (From ANSim Manual, 1989.)

point of light that is turned on causes the nerve cell to get excited. If a light is turned on in the "surround," then the nerve cell becomes inhibited. Turning the light off causes the opposite reactions. There are also "off" center, "on" surround visual fields. Figure 7-16b shows more sophisticated features to which particular nerve cells respond in the cortex. The visual cortex has cells that respond to bars of light and cells that selectively respond to particular orientations or directions of movement. In addition, some cells have been found to respond selectively to more complex visual stimuli, such as the shape of a waving hand.

Figure 7-17 illustrates the analogy between biological feature detectors and feature detecting units in an artificial neural network. The center-surround organization can be used as a feature-detector in the hidden layer of a back-propagating system. It also may, in some cases, be organized by the network itself in response to training data. Clearly, however, the human visual system utilizes a far more complex network than a three- or four-layered back-propagating system.

References

Albus, J. S. 1981. ***Brains, Behavior, and Robotics.*** Peterborough NH: McGraw Hill.

ANSim User's Manual. 1989. Version 1.2, SAIC, 10260 Campus Point Drive, San Diego, California 92121.

Bodian, D. 1967. Neurons, circuits, and neuroglia. In *The Neurosciences: A Study Program*. G. C. Quarton, T. Melnechuk, and R. O. Schmitt, eds. Rockefeller Univ. Press, New York.

Bullock, T. H., Orkand, R., and Grinnell, A. 1977. *Introduction to Nervous Systems*. San Francisco: W. H. Freeman and Company.

Cajal, S. R. 1909–1911. *Histologie du systeme nerveux de l-homme et des vertebres*. (French edition revised and updated by author. Translated from Spanish by L. Azoulay. 2 vols. Maloine, Paris. Republished 1952 by Consejo Superior de Investigaciones Cientificas, Madrid.)

Carlson, N. R. 1977. *Physiology of Behavior*. Boston: Allyn & Bacon.

Cornsweet, T. N. 1970. *Visual Perception*. New York: Academic Press.

Dowling, J. E. 1987. *The Retina*. Cambridge, Massachusetts: Harvard University Press.

Glickstein, M. 1988. The discovery of the visual cortex. *Scientific American*. 259(3):118–127.

Guyton, A. C. 1976. *Organ Physiology Structure and Function of the Nervous System*, 2nd ed. Philadelphia: W. B. Saunders.

Hubel, D. H. & Weisel, T. N. 1979. Brain mechanisms of vision. *Scientific American*. 241(3):150–162.

Kanerva, P. 1988. *Sparse Distributed Memory*. Cambridge, Massachusetts: MIT Press.

Lindsay, P. H. and Norman, D. H. 1977. *Human Information Processing*. New York: Academic Press.

Nieuwenhuys, R. 1969. *The Cerebellum*. In C. A. Fox, vol. 25, Amsterdam: Elsevier.

Sarkisov, S. A., and S. N. Preobrazenskaya, eds. (1959). *Development of the Central Nervous System* (in Russian).

Shepherd, G. M. 1988. *Neurobiology*. New York: Oxford University Press.

Sidman, R. L., and P. Rakic 1982. Development of the Human Central Nervous System, Chapter 1 in *Histology and Histopathology of the Nervous System*. W. Haymaker and R. D. Adams, eds. Springfield, Illinois: C. C. Thomas Pub.

Stotler, W. A. 1953. An experimental study of the cells and connections of the superior olivary complex of the cat. *J. Comp. Neurol.* 98:401–432.

Suga, N. 1978. Specialization of the auditory system for reception and processing of species-specific sounds. *Fed. Proc.* 37:2342–2354.

Thompson, R. F. 1975. *Introduction to Physiological Psychology*. New York: Harper & Row.

8

Biological Synapses

The biological structures that interconnect nerve cells in living tissue have detailed anatomy and physiology that is key to the nature of processing in the nervous system. These synapses and junctions combine electrical phenomena such as membrane potentials with the actions of complex macromolecules. Synapses include anatomical structures so small that they must be viewed at magnifications of 5,000× or more. The architecture of these interconnections determines most of the dynamic activity of neurons as they process incoming signals and generate outgoing signals.

Synapses are key to biological systems in the same sense that interconnections are key to neurocomputing systems. Neurocomputing paradigms rely on interconnection weights and their influence on processing units for the network's computing capabilities. In biological systems, the synapses and junctions between neurons appear to play an equally important role. Hence it is valuable to delve into the detailed structure of interconnections in the one system that solves far more difficult problems than neurocomputing—the human brain.

This chapter provides a mere glimpse into the larger world of biological neural systems. In biological studies, living organisms are examined through experiments that probe anatomical and physiological principles. The operating dynamics of the system is inferred based on indications from experimental evidence. This approach is in sharp contrast to the engineering world where one can describe explicitly how systems such as neurocomputers are built. Biological systems are incompletely understood from the behavioral level down to the level of membranes and molecules. Nevertheless, they are dynamic, complex, and flexible—and they are the most capable neural networks in existence.

CELL MEMBRANES AND IMPULSE PROPAGATION

A cell membrane is a conglomeration of molecules that acts as a containing wall to a cell. Cell membranes have special properties: they can receive and sum signals that arrive from outside the cell and store an electrical potential and propagate impulses down a nerve cell fiber. Membranes sometimes have embedded receptor proteins, which differentiate outside molecules by selective binding, and then cause biochemical changes. The membrane keeps most molecules from passing in or out of the cell, but has special channels for certain ions that allow them to pass through in a controlled way; this passage of ions is significant in the electrical activity of nerve cells, particularly in the integration of incoming signals and the propagation of impulses.

The basic structure of the cell membrane is a phospholipid bilayer, and consists of two layers of phospholipid molecules (Figure 8-1a). These molecules have a phosphoric acid head, which is attracted to water, and a glyceride tail, which is repelled by water. As a result the molecules line up in a water solution in a double layer with their heads directed outward, toward the water, and their tails pointing inward.

The hydrocarbon tails within the membrane pose a barrier to many kinds of molecules. Water molecules pass through, but ions of salts do not, nor do the water-soluble molecules contained within the cell's internal fluid (the cytoplasm). Thus the lipids in the membrane form a natural barrier to retain the cytoplasm. The cell must have special mechanisms for transporting the ions and water-soluble molecules essential for metabolism across the membrane. This mechanism is provided by proteins embedded in the membrane.

The lipids in the cell membrane form a fluid matrix within which a variety of protein molecules are placed. This arrangement is called the ***fluid mosaic membrane model.*** The positions of lipids and proteins are not fixed, they can move laterally across the membrane. Lipids move at a rate of 2 μm/s, and protein molecules move about 40 times more slowly. These protein molecules perform specialized functions, and they may interact with substances inside and outside the cell. Figure 8-1b shows the embedding of complex protein molecules in the phospholipid bilayer.

Membrane proteins play a variety of roles, from structural support to molecular recognition. Some membrane proteins have extremely long sugar chains outside the cell; these are believed to play a role in retaining the cell's shape and in cell recognition (a cell can be recognized by other cells by the size and shape of these exterior chains). Intrinsic membrane proteins lie in the membrane and may span the membrane from one side to the other. Their various functions include reception of neurotransmitters and other molecules that impinge upon the cell. Intrinsic proteins may also aid in movement of molecules across the cell membrane, particularly molecules that are needed for

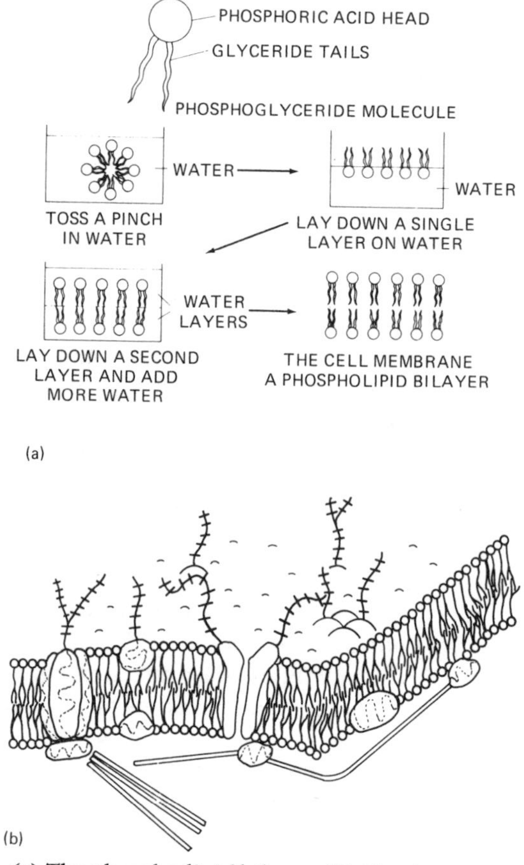

Figure 8-1. (a) The phospho-lipid bilayer. (b) Membrane model with embedded proteins. (From *The Brain: An Introduction to Neuroscience.* By R. F. Thompson. © 1985 by W. H. Freeman and Co. Reprinted with permission.)

cell metabolism, such as glucose and amino acids. Other membrane proteins provide channels for ions to move across the membrane. Some proteins actually pump ions from one side of the membrane to the other, expending energy as they pump. The membrane proteins that provide transport channels from one side of the cell membrane to the other will be described in more detail, as they are important to the electrical activity of nerve cells.

The cell membrane can hold an electrical potential between the inside and the outside of the cell, like a capacitor. A nerve cell typically has a negative potential inside, mostly due to the presence of negatively charged internal proteins that cannot pass through the cell membrane to the outside. The resting state of a nerve cell membrane, its "resting potential," is typically about −70 mV. The resting potential is impressively large considering how small the membrane is—a potential of almost a .10 V is held across a membrane width of less than 1 μ. The power of this storage device can be demonstrated by the electric eel, which taps its membrane potentials to produce charges of hundreds of volts.

Nerve cells have tiny holes through their membranes that allow ions—atoms that have an electric charge—to pass in and out. The most important ions are Na^+, Cl^-, K^+, and Ca^{++}. Each appears to have its own channels through cell membranes. Furthermore, these channels can have "gates" that open and close, which start or stop the flow of ions across the membrane. In a resting state, ions distribute themselves across the membrane, responding to a drive to equalize both their concentrations and the charge across the membrane.

Nerve cell membranes have molecular "pumps" that serve to move ions across the cell membrane. These special proteins are embedded in the cell membrane and expend energy during transport (hence the name "pump"). They force ions to move against their natural equilibrium concentrations,

Possible ion concentrations inside and outside an axon in millimoles (mM) per liter (l) of axoplasm

INSIDE AXON		OUTSIDE AXON	
K^+	= 400	K^+	= 20
Cl^-	= 30	Cl^-	= 590
Na^+	= 60	Na^+	= 436
p^{2-}	= High	p^{2-}	= Very Low

Key: K^+ = Potassium ions;
Cl^- = Chloride ions;
Na^+ = Sodium ions,
p^{2-} = Protein ions,

(a)

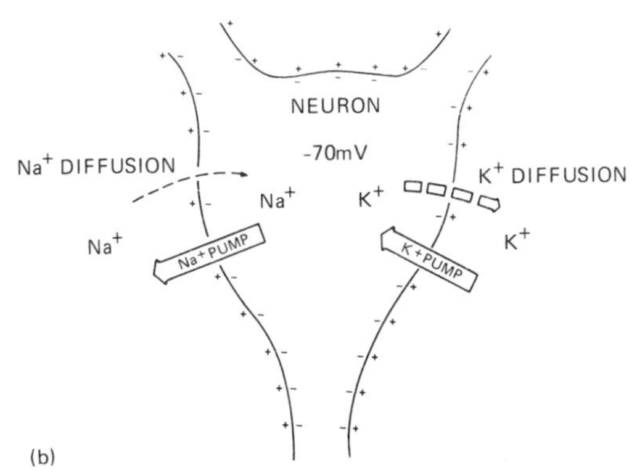

(b)

Figure 8-2. (a) Possible ion concentrations inside and outside an axon. (b) Movements of ions across neuron membrane (Adapted from *Brains, Behavior and Robotics* by J. Albus. McGraw-Hill, 1981).

which requires the expenditure of energy. Pump proteins in the nerve membrane pump Na^+ out of the cell and K^+ into the cell. (Figure 8-2 illustrates the distribution of ions across the cell membrane.) There are more sodium ions on the outside than the inside, and more potassium ions on the inside than on the outside, which is consistent with the directions of pumping actions. The net result of all the ion movements is that the cell maintains a resting potential at about −70 mV.

The resting potential of a nerve cell changes in response to incoming signals from other nerve cells. These incoming signals cause the local membrane potential to be less negative (depolarize) or be more negative (hyperpolarize). The change in potential across the membrane starts at a postsynaptic site, with a transient change in ionic current through channels in the target cell's membrane. The magnitude of the polarization change decays over time, and gets smaller the greater the distance from its point of origin.

Each nerve cell has a threshold value for its membrane potential; if the membrane potential goes above the threshold, then a nerve impulse is triggered. The nerve impulse is usually triggered at the base of the axon, and moves away from the cell soma down to the ends of the axon branches (see Figure 8-3). Carried as a wave of depolarization, the nerve impulse travels very quickly, from a few to over 300 km/h. When an impulse reaches the end of an axon, its arrival causes the release of a chemical substance called a

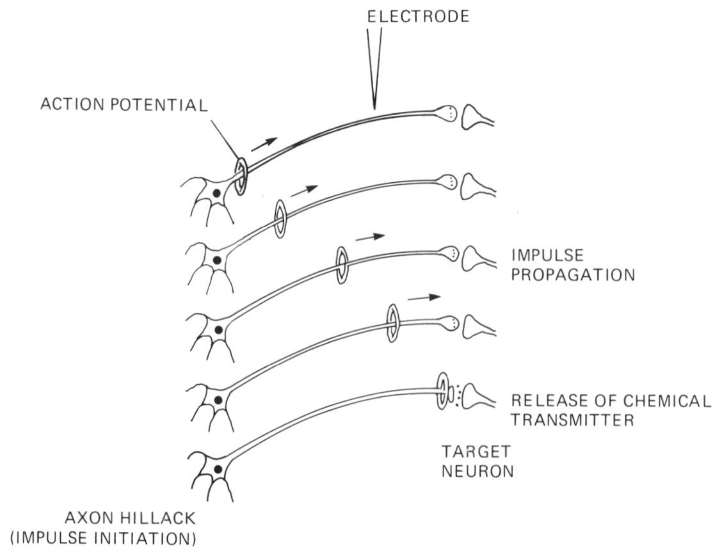

Figure 8-3. Impulse propagating down axon branches (Adapted from The Brain: *An Introduction to Neuroscience.* By R. F. Thompson. © 1985 by W. H. Freeman and Co. Reprinted with permission).

neurotransmitter; the neurotransmitter then crosses the synapse to affect the target neuron.

A small electrode can be placed in the nerve cell to measure the membrane potential during the impulse. Figure 8-4a shows the two parts of the observable waveform: first, the membrane rapidly depolarizes to about +50 mV, then the potential falls, and overshoots the baseline to form the hyperpolarization afterpotential. A slow increase in membrane potential brings it back to its resting value of approximately −70 mV.

The nerve impulse is generated by a sudden exchange of ions across the membrane. This exchange of ions is triggered when the membrane potential reaches its threshold value, which causes a rapid opening of the channels for sodium and potassium. These ions then move quickly toward equilibrium concentrations as sodium rushes in and potassium rushes out. Since the ion channels are membrane proteins that allow passive movement of ions, they do not expend energy for this transfer.

Figure 8-4b illustrates the time course for the opening and closing of the membrane channels. The sodium channels open first and sodium rushes into

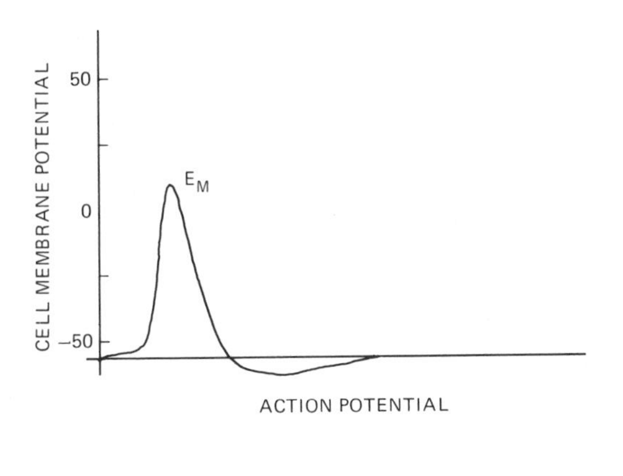

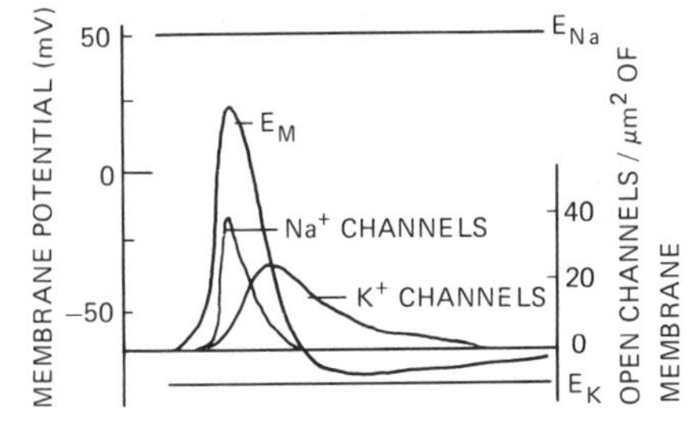

Figure 8-4. (a) Nerve impulse waveform. (b) Opening and closing of membrane channels. (Adapted from *Neurobiology, 2nd Ed.* by G. Shepherd. Oxford Univ. Press. 1988.)

the cell. The sodium is driven into the cell by its concentration gradient (there is more sodium outside, which drives the sodium in), and by the negative charge inside (sodium ions are positively charged, which is attracted to the negative interior of the cell). At the peak of the depolarization, the sodium channels start closing and the sodium influx declines. The membrane potential soon begins to drop.

The potassium channels open on a slower time course, as shown in Figure 8-4b. Potassium ions move out of the cell as the membrane potential returns to its resting level, becoming more negative inside. Potassium channels start to close at about the same time that the membrane reaches its resting potential. Potassium ions then participate in the brief hyperpolarization afterpotential at the end of the impulse, which is followed by a final return back to resting potential.

Nerve cells have a refractory period after firing an impulse. During the refractory period, the axon cannot be stimulated to generate another action potential. The refractory period results from the ions and proteins not having returned to their initial state of readiness to fire a new impulse. Generally the refractory period lasts about 3 – 5 ms after an action potential. This waiting period causes a requisite lag between successive impulses from the same nerve cell; thus there is a maximum firing rate that the nerve cell cannot exceed.

A fine microelectrode can be ground and placed in the vicinity of an active nerve cell, and the impulses that go down the axon can then be displayed on an oscilloscope. A time record of these impulses is called a ***nerve impulse train*** or ***spike train*** (see Figure 8-5). The waveforms of successive impulses appear the same; thus, the shape of the impulse is not important. The important aspects of the impulse train are the times of occurrences of impulses, the times of occurrence of volleys of impulses, and the temporal delays between successive impulses. ***Which*** nerve cells fire also appears to be important in the nervous system as well as groups of nerve cells that tend to fire together or in particular types of patterns (Dayhoff 1987; Dayhoff and Gerstein 1983; Abeles and Gerstein 1989; Gerstein, Perkel and Dayhoff 1985; Lindsey and Gerstein 1979a,b; Lindsey et al. 1989).

Some nerve cell axons are insulated by myelin sheaths, which consist of glial cells that wrap themselves about a nerve axon. Myelin can speed the conduction of impulses considerably, and is especially important for long axons such as motor axons, which must reach all the way to the muscles they are to stimulate. A myelin sheath leaves gaps, or nodes, which expose segments of bare membrane. At these nodes, the ion exchanges for the nerve impulse occurs; the impulse transmission travels more rapidly between nodes than in the unmyelinated fibers (see Figure 8-6).

Traditionally, a nerve cell that transmits impulses down its axons and re-

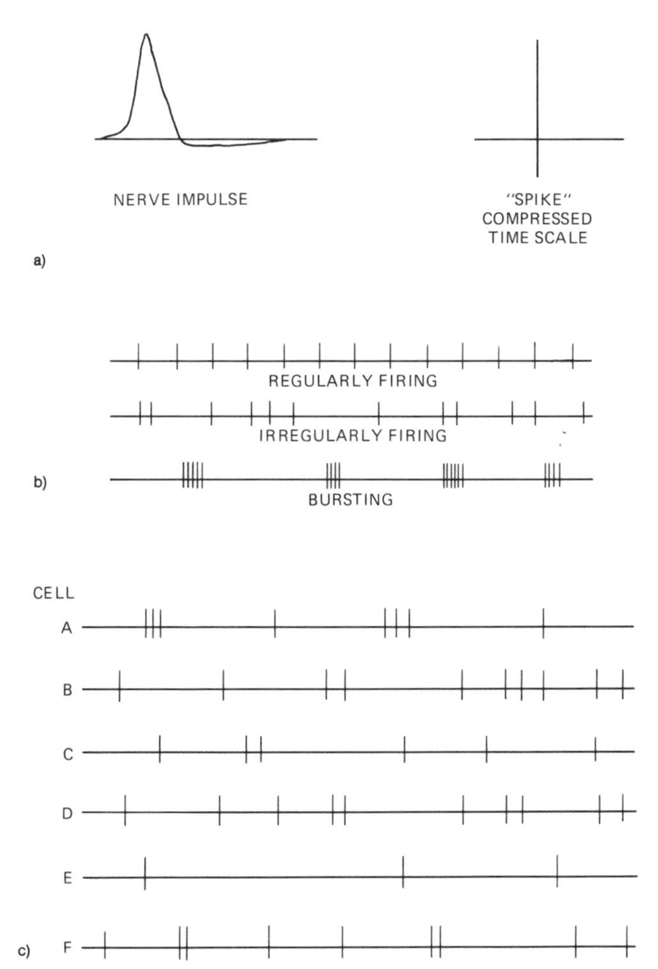

Figure 8-5. Nerve impulse trains, based on hypothetical data. (a) The recorded impulse waveform is compressed to appear as a spike. (b) A regularly firing impulse train, an irregularly firing impulse train, and a bursting neuron. (c) Simultaneously recorded impulse trains from six different neurons.

ceives impulses at its dendrites has been considered a standard neuron, however many exceptions to this "standard" exist. Not all nerve cells generate impulses, for example those in the retina. Those that do not generate impulses rely on their ability to spread changes in membrane potentials along the cell membrane. These changes can cause the release of neurotransmitter at synapses, and thus affect target cells without the use of impulses. Occasionally a nonnerve cell has been known to generate impulses, as in the Venus flytrap, which uses impulses to respond rapidly to movements of prey.

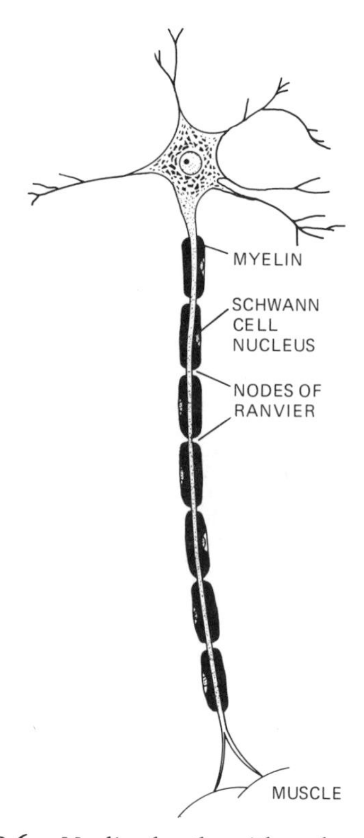

Figure 8-6. Myelin sheaths with nodes of Ranvier.

The cell membrane plays a very important role in nerve cell activities. It receives signals across synapses, sums arriving signals, and spreads impulses down the axon. The membrane poses a barrier to ions, which can be exchanged across the membrane during membrane activity and impulse propagation. The diversity and capability of membrane proteins is a powerful building block for the complex behaviors of nerve cells, cell assemblies, and biological nerve networks.

SYNAPSES—THE BASIC STRUCTURE

Figure 8-7a, a schematic diagram of a synapse, includes the presynaptic cell, its synaptic vesicles, the synaptic cleft, and the postsynaptic cell. In the presynaptic cell, arriving action potentials change the membrane in the presynaptic area, and cause a release of a transmitter from the synaptic vesicles. The transmitter crosses the synaptic cleft to reach the postsynaptic cell, where

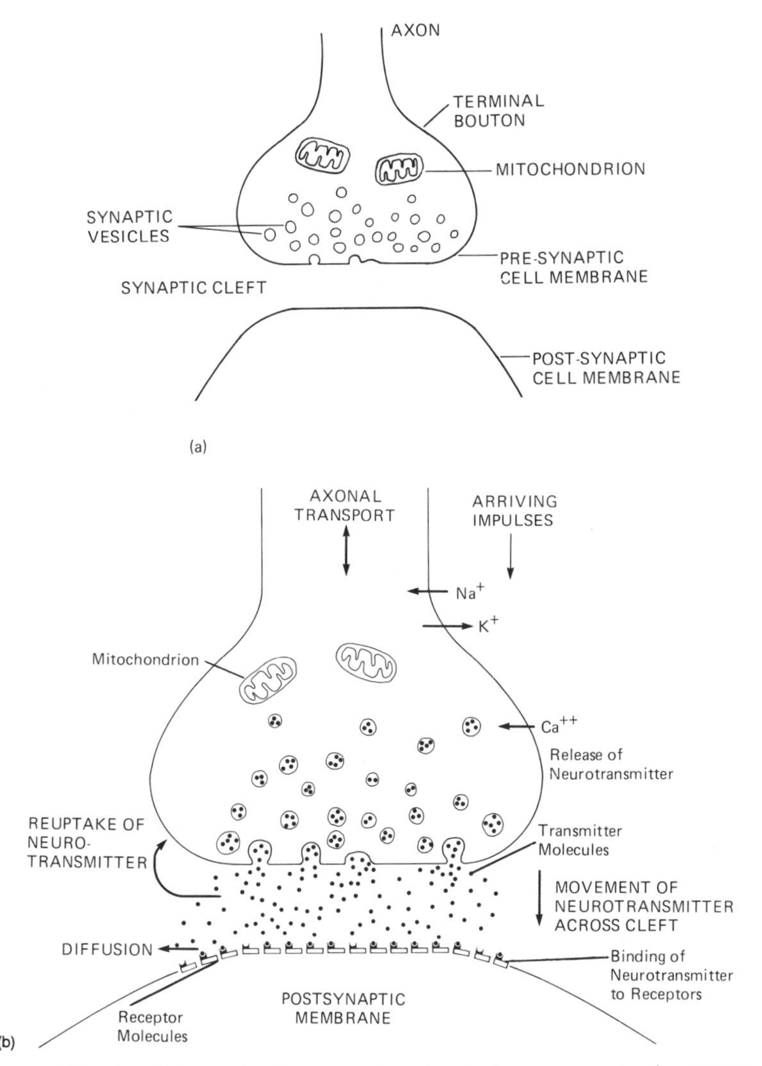

Figure 8-7. (a) Schematic diagram of anatomical structures in the synapse. (b) Dynamic processes of the synapse.

it influences the membrane properties of the receiving cell. The likelihood of the postsynaptic cell firing is then changed—either increased or decreased.

The presynaptic cell usually has a bouton at the end of its axonal process. A cell may have many such boutons, as there can be many branches from a single axon. An impulse usually propagates simultaneously down each branch, although sometimes impulses are shunted to one branch and not to another

(Spiral et al. 1976). The arrival of the action potential affects the presynaptic cell membrane at the bouton terminal; this membrane responds by allowing an influx of calcium (Ca^{++}) ions. Next, the transmitter is released from the synaptic vesicles into the cleft. The release of the transmitter appears to be modulated by calcium ions; calcium ions are involved in candidate mechanisms for the modulation of synaptic strength during learning (Brown et al. 1988).

The synaptic vesicles are concentrated at the synaptic area of the presynaptic cell. Usually these are clustered into a dense group, but sometimes there is a more regular formation (Figure 8-8). The vesicles are filled with a neurotransmitter substance and are encompassed by their own membrane. When the synaptic vesicles release the neurotransmitter, their membranes fuse with the cell membrane in a process known as *exocytosis,* and the internal fluid is released into the extracellular space.

Synaptic vesicles come in different shapes, and may contain any one of a number of different neurotransmitters. The appearance of their shapes (after preparation) ranges from spherical to ellipsoidal or flattened. Although these are sometimes classified into two different classes by shape, there is not a sharp distinction. Usually the vesicles in a synapse all contain the same transmitter substance, although exceptions exist (Millhorn and Hökfelt 1988).

Transmitter substances may have an excitatory or inhibitory effect. Many of the excitatory synapses use glutamate as a transmitter; many inhibitory synapses use GABA as a transmitter. For these synapses, glutamate has a depolarizing effect on the postsynaptic membrane, which increases the chance of firing for the target neuron, since depolarization moves closer to the threshold potential. GABA has a hyperpolarizing effect on the postsynaptic membrane, which decreases the chance of firing for the target neuron, since hyperpolarization moves farther away from the threshold potential. A variety of other transmitters are found, including acetylcholine and other amino acids and peptides (short proteins).

Transmitter substances, when released, move across the synaptic cleft by diffusion and their movement is influenced by the cleft extracellular matrix, the fluid that contains ions and organic molecules. Figure 8-7b diagrams the biochemical processes in the synaptic cleft. Vesicles empty into the cleft, and transmitter molecules move across. Most molecules reach the postsynaptic neuron, where they bind to the receptor molecules on the postsynaptic cell membrane. The excess transmitter molecules are removed through diffusion or degradation by enzymes, or are taken up for reuse by the presynaptic cell. New neurotransmitter molecules are synthesized by the presynaptic cell and transported down the axon to the terminal bouton.

The postsynaptic cell has a membrane region dense with receptor molecules. Receptor molecules use a "lock and key" mechanism — only a specific

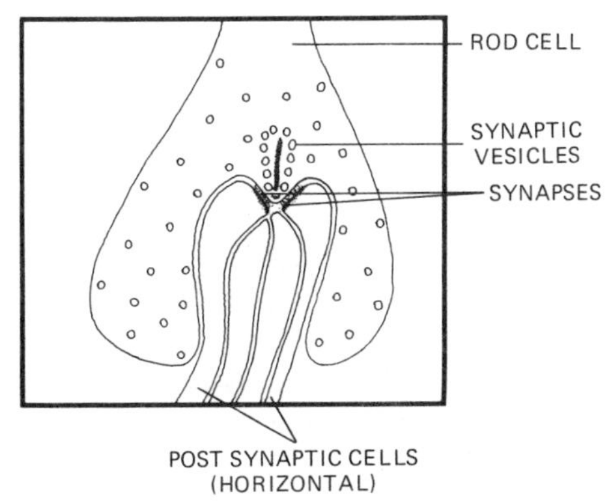

Figure 8-8. A synapse with organized presynaptic vesicles. (From *The Retina* by J. Dowling © 1987. Reprinted by permission of the Belkhap Press of Harvard Univ. Press.)

substance can fit with a specific type of receptor to make binding possible—when binding to substances in the cleft. When a molecule binds to a receptor, the configuration of the receptor molecule is changed. In turn this change influences the postsynaptic membrane's properties. Thus it is the properties of the receptor and not the transmitter molecule that ultimately defines the action of a transmitter.

There are two types of membrane receptors. For the first type of receptor protein, binding of the neurotransmitter molecule induces a conformational change in the protein's three-dimensional shape, which opens up a channel through the membrane to allow ions to flow. The second type of receptor molecule is a nonchannel protein, for which binding of the neurotransmitter causes further reactions with nearby enzymes and proteins. These reactions ultimately affect the postsynaptic membrane properties. Receptor mechanisms thus allow for a wide range of brief or long-lasting effects on the postsynaptic terminal and neuron.

Figure 8-9 shows a model of a three-dimensional receptor molecule that does selective binding to the neurotransmitter acetylcholine (ACh) (Kistler et al. 1982). This transmitter is typically used in synapses from motor neurons to target muscle cells, to initiate muscular movements. There is a portion of the molecule outside the membrane, which binds selectively to ACh. Thus, binding to ACh—but not other substances—is possible in the "lock-and-key" model. There is another portion of the receptor molecule in the membrane: Its subunits come together in a circular arrangement, which forms a channel, or "central pore." This channel acts as a funnel, allowing ions to pass through. The binding of ACh to the outer portion of the receptor molecule brings about a conformational change which allows the channel to open up

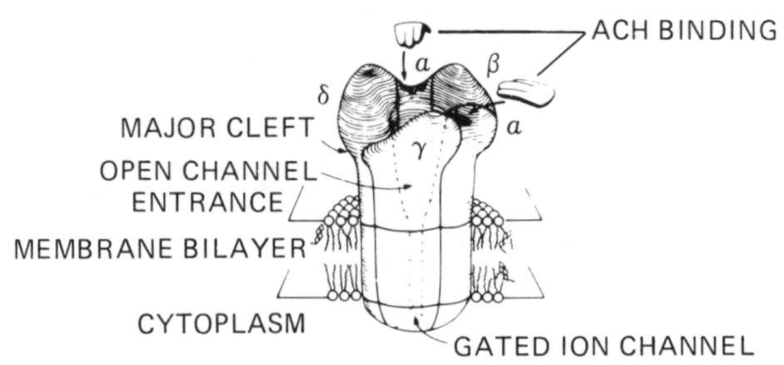

Figure 8-9. Three-dimensional model of the ACh receptor, showing arrangement of subunits around a central channel. (Adapted from Kistler et al. *Biophysical Journal.* 1982. Rockefeller Press.)

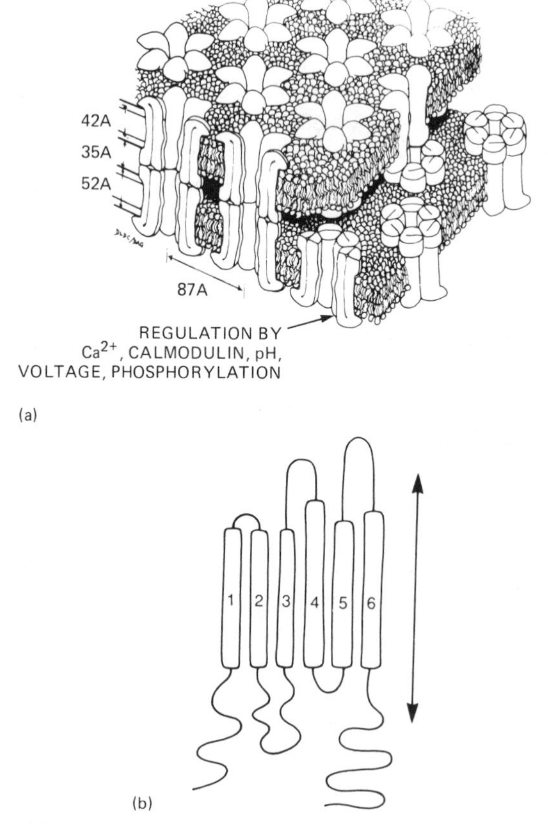

Figure 8-10. (a) A model for gap junction structure. (b) Gap junction protein. (From Makowski, Caspar, Phillips, and Goodenough. Gap junction structures II. Analysis of the x-ray diffraction data. *J. of Cell Biology.* 1977. Rockefeller P.)

and pass ions through. When acetylcholine is not bound to the receptor molecule, the pore is closed.

Figure 8-10 shows a model for another type of junction between neurons —the gap junction (Makowski 1977). A gap junction typically has only a 2–4 nm separation between the membranes of the adjacent neurons. This distance is much closer than the synapses described earlier. Gap junctions provide low-resistance electrical interactions between two cells. They form cell-to-cell channels for ions to pass directly from inside one cell to inside another. Gap junctions are more common in primitive animals such as invertebrates and lower vertebrates than they are in the mammalian brain, leading some to believe they may be a primitive form of synapse.

Dendritic spines (shown in Figure 8-11) appear key to successful interconnections in biological systems. The spines, which cover dendritic trees, provide sites for synapses, and the regulation of spine size and shape may control synaptic efficacy (Perkel and Perkel 1985). In certain types of mental retardation, dendritic spines are missing, as shown in the figure, indicating that they play a necessary role in healthy brain interconnections.

Biological synapses have an elaborately detailed anatomy, including a presynaptic cell, synaptic vesicles, synaptic cleft, and postsynaptic cell. Their activities include the release of a transmitter in response to arriving impulses,

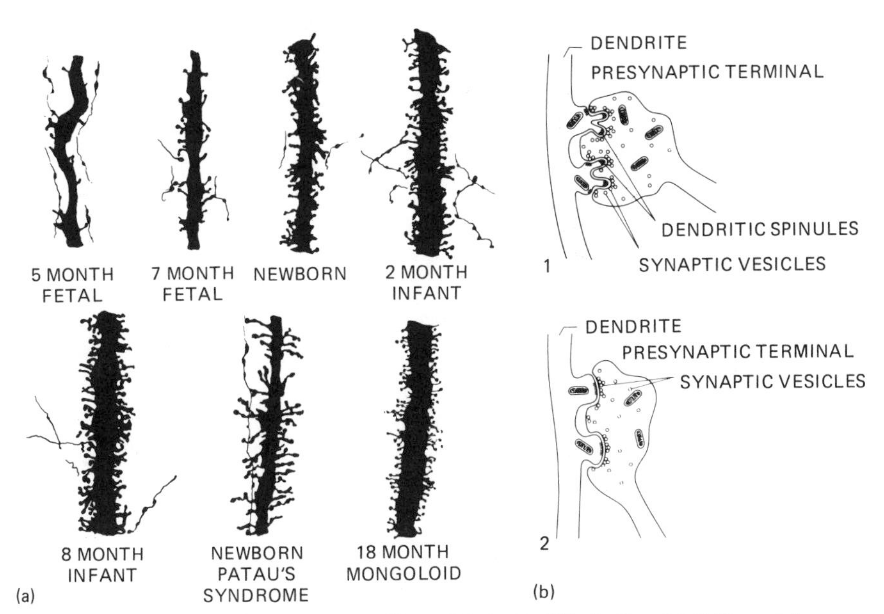

Figure 8-11. Dendritic spines. (From Marin-Padilla in Lund 1979; and from Hamori in Hofer 1981.)

the modulation of that release to influence the strength or nature of the connection, the movement of a transmitter across the cleft, and the binding of transmitter substances onto the postsynaptic membrane. An alternative type of connection is found in the gap junction, which may be a primitive form of synapse. Synapses form the basic interconnecting link among neurons, and are therefore the quintessential building block of biological neural systems.

THE VARIETY OF SYNAPSES AND JUNCTIONS

Biological systems use a wide variety of synapses and junctions in their intercellular communications. A panoramic view of this variation shows the underlying complexity of the interconnections in biological neural systems and enables us to compare biological systems to artificial neural networks.

Biological synapses differ in the cell parts that participate in the synaptic connection (e.g., cell dendrites, soma, axons). The typical synapse described in textbooks is axo-dendritic, a contact made from an axon to a dendrite. An axo-dendritic synapse is illustrated in Figure 8-12, as is a variety of other types

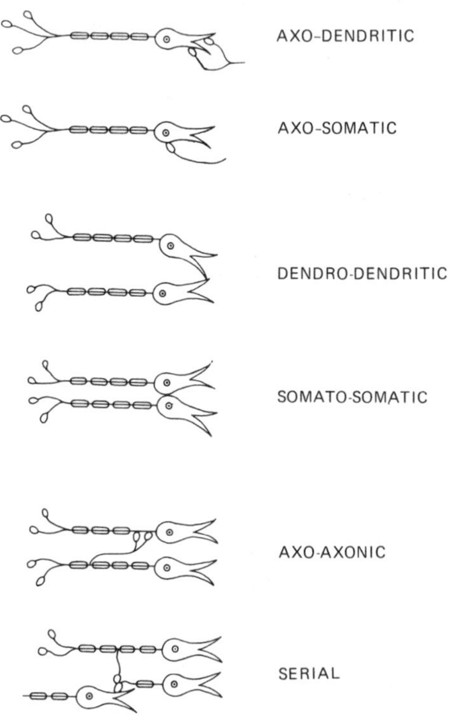

Figure 8-12. Different types of synaptic connections (from D. Bodian. Neuron junctions. *Anatomical Record* Alan R. Liss 1972).

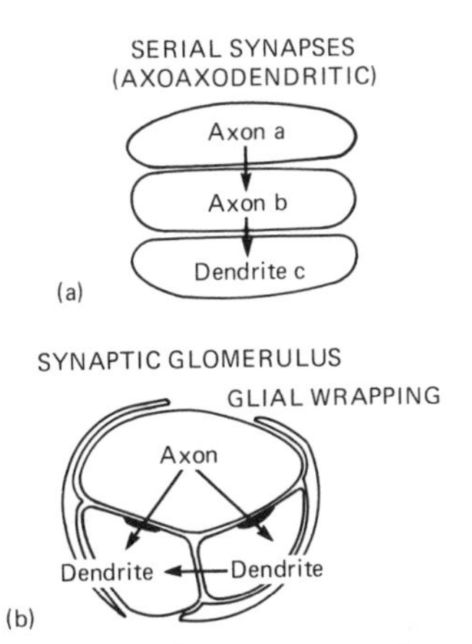

Figure 8-13. (a) A serial synapse. (b) A synaptic glomerulus. (Adapted from *Neurobiology, 2nd Ed.* by G. Shepherd Oxford Univ. Press. 1988.)

of frequently observed synaptic connections. An axo-somatic synapse, going from an axon to a cell body, is shown, as well as a dendro-dendritic synapse, which makes contact from one dendrite to another. Other synapses go from axon to axon (axo-axonic) and from cell body to cell body (somato-somatic). Sometimes serial or clustered synapses occur in close proximity (bottom of Figure 8-12). We do not know as much about the detailed dynamics of these other synapses as we do about the "standard" axo-dendritic synapse.

Often there are two or more synapses situated near each other. Figure 8-13 gives two such examples: (a) shows an axon that synapses to another axon, which in turn synapses to a dendrite. These synapses are arranged in a serial fashion. An alternative arrangement, with synapses in parallel, is shown in (b), where an axon synapses to two dendrites. Arriving pulses cause release of neurotransmitter substances at both of these synapses simultaneously. In addition, the two dendrites have a synapse between them.

Figure 8-14 shows more synaptic structures and topography. Ordinary synapses with three types of vesicles are shown: spheroid, flat, and dense core vesicles. A gap junction is also shown, with a narrower distance between the participating cells, and no synaptic vesicles. Junctions with two different types of contacts between the same two cells are also possible. A mixed junction, with a synapse and a gap junction, is shown on the right. A reciprocal junction is shown, with two cells that each synapse on the other cell. Reciprocal junctions are surprisingly common. A simple synaptic cluster, with two synapses and three cells, is shown on the bottom right.

Tightly grouped clusters of synaptic terminals are called synaptic glomer-

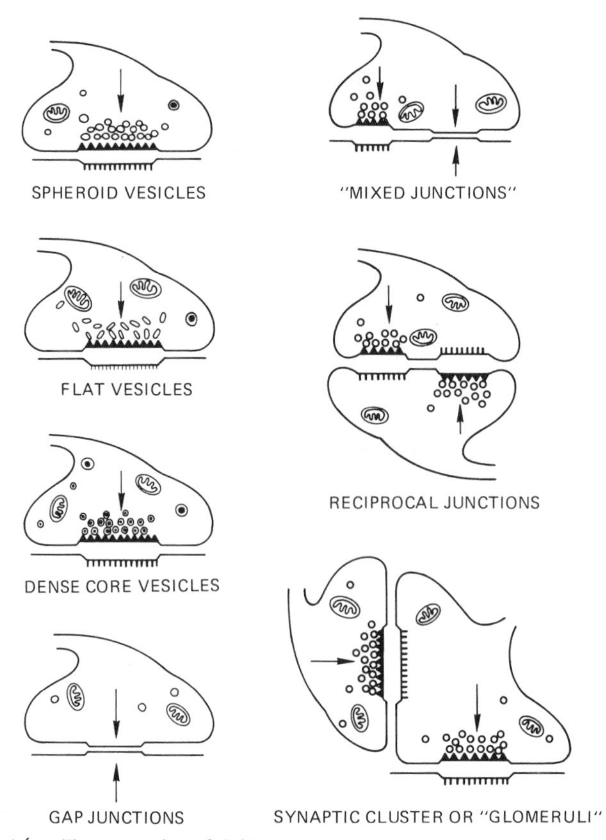

Figure 8-14. Topography of different synapses (from D. Bodian. Neuron junctions. *Anatomical Record.* Alan R. Liss 1972).

uli. The synaptic cluster at the bottom right of Figure 8-14 is considered a glomerulus. Figure 8-15 shows two glomeruli that are more complex. The upper diagram, from a neural circuit in the cerebellum, consists of a single axon surrounded by dendrites. There are other axon terminals among the dendritic areas. Clusters of vesicles indicate where the many synapses are located. The lower diagram in Figure 8-15 is taken from the thalamus and shows a single dendritic spine flanked by arriving axons. The axons synapse to each other as well as to the dendrite, and there are many synapses that are dense with vesicles.

The variety in synapses present in biological systems is important to consider in the analysis of neural systems. It is striking to biological scientists that so much variety exists; sometimes it is difficult to identify the regularities in a

DIVERGENCE

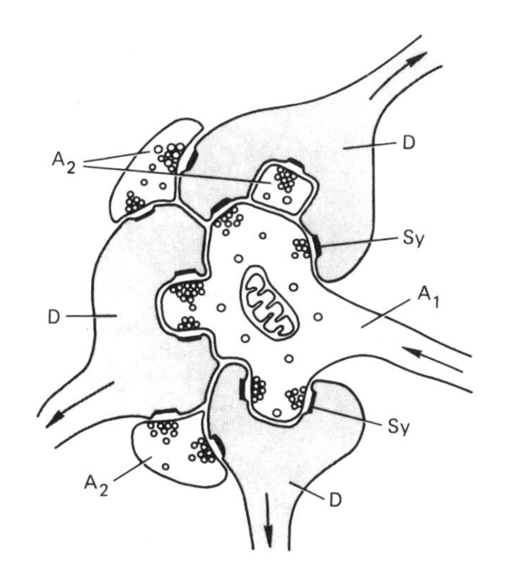

CONVERGENCE

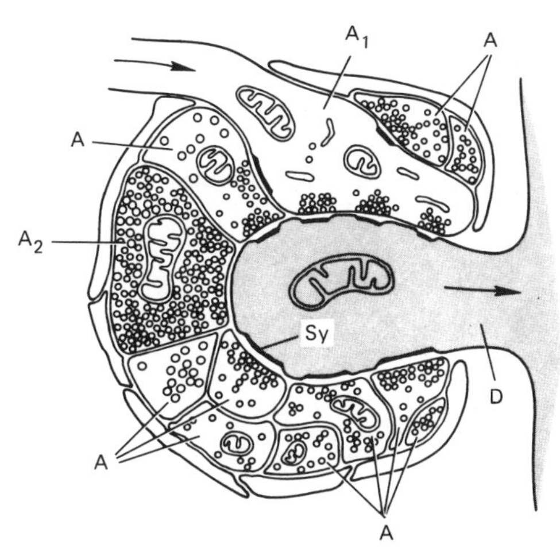

Figure 8-15. Synaptic glomeruli. (a) Divergent glomerulus, from the cerebellum. (b) Convergent glomerulus, from the thalamus. (From U. Steiger, Über den Feinbau des Neuropolis im Corpus pedunculatum der Waldameise. *Zeitschrift Zellforschung*. 1967. Springer-Verlag.)

biological system because of all of the variation that can be observed. David Bodian (1972), a scientist who catalogued synaptic structures, stated:

> In synaptic systems . . . we see not a stereotyped mechanism for the transfer of information from cell to cell, but another display of the fact that every conceivable capability of living organisms to solve adaptive problems is likely to be put to the test in the evolution of life.

As yet we do not know the role that much of the synaptic variety plays in the nervous system, and have not evaluated most of the biological structures here for possible inclusion in artificial neural systems. The mechanisms of evolution that put these structures in place are unique, and are very different from the way structures are chosen for inclusion in artificial neural net models.

POSTSYNAPTIC SUMMATION

The most important activity of single neurons is summation and integration of arriving inputs. Summation in biological units is different from that in artificial neural networks. Artificial networks sum arriving scalar values at discrete

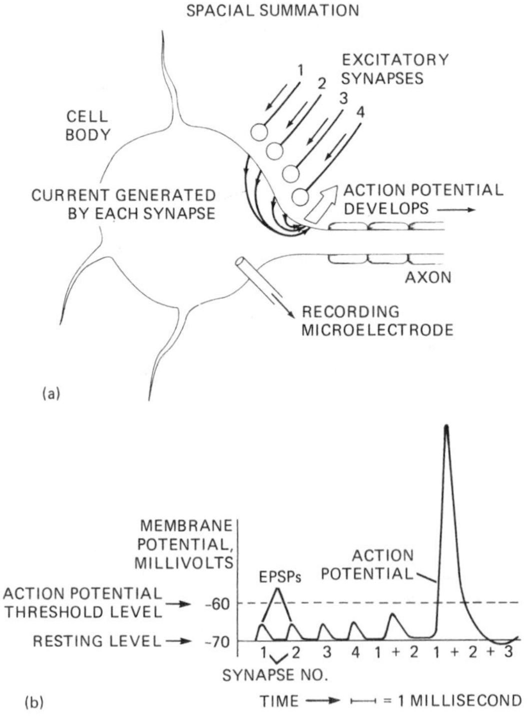

Figure 8-16. Summation of arriving signals. (a) Spike initiation zone. (b) An individual EPSP.

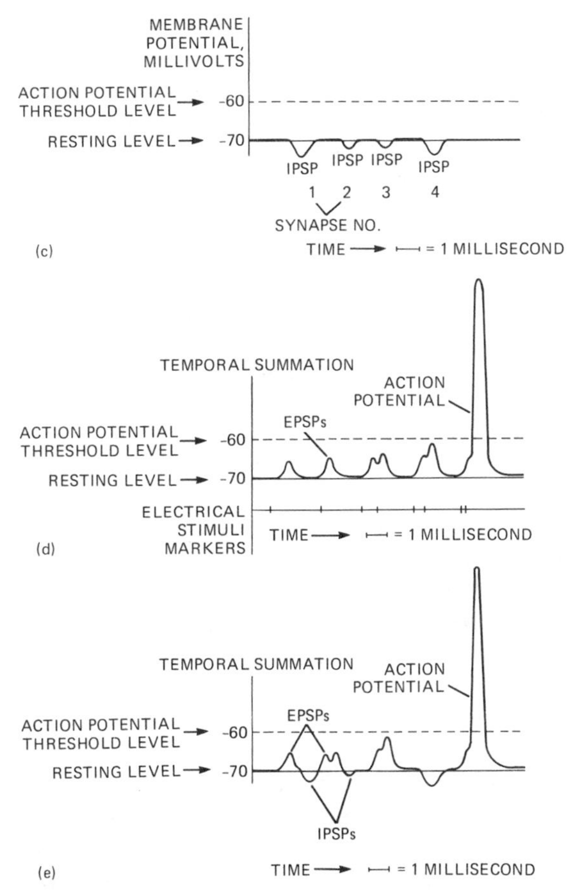

Figure 8-16. (cont.) (c) An individual IPSP. (d) A summation of arriving EPSPs. (e) A summation of both EPSPs and IPSPs. (Adapted from The Brain: An Introduction to Neuroscience. By R. F. Thompson. © 1985 by W. H. Freeman and Co. Reprinted with permission.)

time steps. Biological neural networks have mechanisms for temporal integration as well as summation of incoming signals, and biological mechanisms for inhibition are different from those for excitation, producing effects with differing magnitudes and temporal dynamics.

When a neurotransmitter is received at a postsynaptic cell, the cell's membrane is affected. The way in which this effect occurs is key to how the target neuron sums its inputs. The process is time modulated, spacially modulated, and influenced by the conductances and locations of ions.

When an excitatory synapse is activated, and a transmitter released, the postsynaptic neuron typically generates an excitatory postsynaptic potential (EPSP), an example of which is shown on the top of Figure 8-16a,b. There are two parts to the EPSP response: first, the rapid rise in membrane potential as a result of the arrival of the neurotransmitter and its action on the postsynaptic receptors, and then a temporal decay when the membrane potential returns to its resting value over time. The EPSP is spread spatially over the neuron — its height usually decays over a distance from the postsynaptic site. When the membrane changes from an EPSP reach the spike initiation zone, they influence the timing of when the next impulses are generated.

An inhibitory postsynaptic potential (IPSP) is also possible, typically in response to a transmitter (or synapse) that has an inhibitory influence. Figure 8-16c shows an IPSP in the postsynaptic membrane potential. There is a small initial drop, with a decay over time. The initial drop is sometimes smaller in magnitude than the initial rise in an EPSP, and the decay time may be different than for an EPSP. Typically the duration of the postsynaptic membrane change is key to the effectiveness of the inhibition. Because of different parameters and mechanisms, excitation and inhibition are not symmetric in the nervous system.

EPSPs and IPSPs can be summed by the target neuron. Figure 8-16d shows the summation of rapidly arriving EPSPs, and Figure 8-16e includes effects of arriving IPSPs also. EPSPs make the membrane potential less negative, and IPSPs tend to lower or hold the membrane potential at subthreshold levels. Sometimes these two types of changes can be seen distinctly in the membrane potential trace.

Yet another factor enters into the summation of incoming impulses — the size and shape of the dendritic tree and its processes. The dendrite branching topology, the branch diameters, and the trunk size of the dendritic tree all influence the summation of incoming signals. The position of the synapse on the dendritic tree or on the cell body also influences how arriving pulses are summed. Extensive modeling studies on these types of relationships have been done by Rall (1978) and others (MacGregor 1987).

A pivotal issue is identifying what makes neurons fire. Usually a number of incoming impulses are required to make a target cell fire. If the arriving impulses are spaced closely in time, then the target cell sums the EPSPs rapidly, with little time to decay between arrivals. Thus, impulses that are closely spaced in time are more likely to cause the target neuron to fire. These incoming impulses could be from the same or different neurons. A group of neurons that all feed the same target neuron could trigger a response if they fire in near synchrony.

A number of studies have been performed on groups of neurons that tend to fire together. Statistical methods have been developed to examine impulse train data recorded simultaneously from multiple neurons, to identify groups of neurons that tend to fire together, and to define their functional connectivity and dynamics (Gerstein, Perkel, and Dayhoff 1985). Temporal synchrony has been found under a variety of conditions. The correlations in the firing times of multiple neurons can change spontaneously and can change in response to stimulation. For example, information about spatial direction of a sound source is available in the near-coincident firings of neurons even though it is not present in the spike trains of the individual neurons (Aertsen and Gerstein 1985).

A question remains about the possible significance of temporal patterns in the timing of pulses from neurons and assemblies of neurons. Favored patterns—those that repeat unexpectedly—have been found (Dayhoff and Gerstein 1983; Abeles and Gerstein 1988), and other temporal patterns have appeared in a variety of preparations (Richmond et al. 1987). Such patterns might have biological significance in contributing to the dynamic processing and encoding done in biological neural networks. Although a temporal pattern is not expected to elicit an individual pulse from an individual target neuron, such a pattern contributes to temporal patterns that are output from assemblies of neurons. At the least, this patterning could be a side effect of the way the nervous system processes. At most, impulse patterns and correlations could be a code for information processing and representation. It is possible that identifying and tracking assembly patterns could turn out to be a highly significant way of approaching and understanding neural dynamics.

The structure of biological neural networks is at once both divergent and convergent. The output of a neuron is fractionated, as its axon branches to synapse onto many other neurons. A single neuron in turn receives input from many sources. The temporal spacing of impulses is superimposed on this anatomical structure of divergent and convergent interconnections. Biological nervous systems, thus, can be considered to perform *spatiotemporal processing,* a much more complex process than the parallel or serial processing of conventional computers.

COMPARISON WITH NEUROCOMPUTING

An overview of neural networks is not complete without a discussion of how the artificial architectures deviate from the biological. Synapses and interconnection strengths are a focal point in both systems and are key to their differences. Other differences arise in considering properties of entire networks or assemblies of units—emergent properties such as redundancy, pattern recognition, and learning (see Table 8-1 for a summary of these differences).

The most fundamental difference between biological systems and artificial neural networks is in the complexity of the synapses. These interconnection

TABLE 8-1. A COMPARISON OF BIOLOGICAL AND ARTIFICIAL NEURAL NETWORKS.

Biological Neural Networks	Artificial Neural Networks
Synapses complex	Synapses simple
Fixed gross wiring structure plus variation in detailed structure	Usually fully interconnected slabs
Pulse transmission	Activity value and connection strengths
Topological mappings	Kohonen feature map
Distributed representations and processing	Distributed representations and processing
Redundancy	Redundancy
Feature detectors	Feature detectors
Learning as fast as one pass	Slow to converge
100 billion neurons	Usually up to hundreds or thousands of neurons
Estimated 10,000 interconnections per neuron	Usually 10–10,000 interconnects per neuron
Continuous or asynchronous updating	Generally synchronous updating

points have myriad components and active processes in biological systems. The strength of biological synapses may be affected by any of the following:

1. Size and number of synaptic vesicles
2. Content of synaptic vesicles
3. Amount of neurotransmitter released from each arriving impulse
4. Release and uptake rates of neurotransmitters
5. Rate of formation of synaptic vesicles
6. Influx and outflow of ions: Ca^{++}, Na^{++}, K^{+}, Cl^{-}
7. Binding strength for neurotransmitter at postsynaptic site
8. The formation and disappearance of the synapse itself, in response to environmental or internal stimuli

In principle, any of these items may vary or affect the synaptic strength in a biological system. Research is currently underway to identify those specific mechanisms that are used to modulate synaptic strength in nervous systems.

In addition, biological systems can have connections that are stronger or weaker as a result of structure larger than individual synapses. For example, the number of synapses on a particular cell, the number of axon branches, the topology and size of the dendritic branches, and the location of the synapses on the dendritic tree all affect the strengths of interconnections among nerve cells. There could be two or more secondary synapses affecting a primary synapse (glomeruli). The physical placement of axons and dendrites can enhance a connection, as in the climbing fibers, where the axon branches match the dendritic tree so that each branch may synapse many times.

In contrast, artificial neural networks have relatively simple interconnections. For example, summation in standard back-propagation can be expressed with a simple linear weighted sum combined with a nonlinear sigmoid (threshold) function. Other simple calculations may be done, such as inhibition and competition. Sometimes additional terms are added to the summation functions, such as a dependency on past changes, or the addition of second-order terms (Rumelhart et al. 1986; Parker 1987). However, these do not rival the biological synaptic processes in complexity.

Biological systems send signals from one unit to another by means of impulse transmission. Artificial neural networks can transfer precise scalar values from unit to unit. Biological impulse transmission takes place at any time and its timing is determined by incoming signals. Artificial networks update their parameters periodically, in discrete time steps. Usually the whole network is updated at the same time (in effect). Thus artificial networks can be considered to have synchronous updating, whereas in biological systems the updating is asynchronous. Biological neurons update whenever an impulse arrives, and may also have parameters that decay or change between arrivals.

One advantage of artificial networks is that the scalar values that are transferred from unit to unit can be implemented to be relatively precise. In biological systems, a single interconnection does not transfer a precise scalar value. If average firing rate is considered to be the value transferred, then it has limited precision, especially over a short time frame. Biological systems, on the other hand, have a built-in temporal structure because impulses can occur at any time, and thus may form temporal patterns. The summation in the two systems must be done differently as a result of their different signaling characteristics.

Biological systems have predetermined wiring at the system level. For example, the major fibers and connections in the visual and auditory systems are the same for different individuals (see Chapter 7); at a more detailed level the circuitry appears to be different for different individuals. For example, isogenetic animals (animals with exactly the same genes) do not have corresponding neurons with the same dendritic branching structure and topology; it may not even be possible to find individual corresponding neurons. (Thus there is not a one-to-one correspondence of neurons in the brain from individual to individual.) Concomitantly, no synapse-for-synapse correspondence, nor an assembly-for-assembly correspondence has been identified.

Artificial neural networks are usually layered and fully interconnected, with all units in a given layer connected to all units in the layers above and below. Artificial networks can also be sparsely interconnected, or have connections removed selectively after training. Biological systems have layered structures: The cortex, geniculate, cerebellum, and hypocampus all have elegant layered organization. However, the layers of biological systems are not simple rows of independent units, as in most artificial networks, and the neurons in each layer or cluster tend to be densely interconnected to one another, again in contrast to most artificial networks. The layers in biological systems are not fully interconnected with layers above and below in the simplistic way that is found in many artificial systems, as biological connections may be sparse or may involve more than one synapse. Three-dimensional packing considerations discourage fully interconnected topologies for biological systems, and also pose a constraint in designing neurocomputing hardware.

Feature detectors occur in both biological and artificial systems. Back-propagation organizes feature detectors in the middle layers of a system with three or more layers (see Figure 4-14). Biological systems appear to do an abstraction of special features from sensory inputs. Cells in the mammalian cortex, for example, strongly prefer certain visual or acoustic patterns.

Distributed representations and distributed processing occurs both in artificial and biological networks. A pattern mapping, for example, uses the *entire* network, not just a single location in the network, to determine its output pattern given the input pattern. Each pattern mapping uses the whole network, and the same connection strengths determine the correct output for many different input patterns. The information needed to do this mapping is thus distributed throughout the network. Distributed representations and processing is a characteristic of both biological neural systems and artificial neural networks.

Redundancy occurs both in artificial neural networks and in biological networks, and is an important characteristic of both systems. Redundancy can increase the reliability of a system, allowing it to function even when some of the neural units are destroyed. Back-propagation naturally organizes redundancy into a network if the network has enough units (i.e., the network needs an excess of internal units to implement redundancy). Artificial neural networks are of interest because their reliability in the face of damage may be critical for many applications, such as defense systems and medical systems.

In comparison, we know that biological networks use redundancy because when some cells die, the nervous system still appears to perform as before. Studies have shown that overlapping receptive fields are a type of redundancy usually built into mammalian sensory systems. In addition to providing reliable performance in the face of cellular damage, redundancy may also counteract sources of noise in biological systems.

The human brain is extremely large for a neural network, containing 10^{11} neurons, with as many as 10^{4} interconnections each. This leads to a tremendous number of possible neural assemblies and activity patterns that could, in principle, be used in neural function. Artificial neural systems, in contrast, have usually been limited in studies to under 10,000 units, with perhaps hundreds of connections per unit. Scale-up properties of artificial neural networks are as yet unknown, and will be the subject of future investigations.

Biological systems have the remarkable property of being able to learn in as little as one training presentation. A face that is viewed once, for example, can be recognized again. In contrast, artificial systems are usually slow to converge, and usually require hundreds or thousands of training presentations in order for learning to take place. Artificial systems focus on learning a single task; biological systems have broad capabilities and can address many different types of tasks. Evolution has had a long time to work out the details of biological systems, utilized different methodologies, and had different requirements compared to today's builder of artificial neural systems. While engineers and scientists struggle to prove the technology of artificial neural networks, the success of evolution is indisputable.

References

Abeles, M. and G. L. Gerstein. 1989. Detecting spaciotemporal firing patterns among simultaneously recorded neurons. *J. Neurophysiol.* 60:909–924.

Aertsen, A. and Gerstein, G. L. 1985. Evaluation of neuronal connectivity: Sensitivity of cross correlation. *Brain Res.* 340:341–354.

Albus, J. S. 1981. *Brains, Behavior, and Robotics.* Peterborough, N.H.: McGraw Hill.

Bodian, D. 1972. Neuron junctions: A revolutionary decade. *Anat. Rec.* 174:73–82.

Brown, T. H., Chapman, P. F., Kairiss, E. W., and Keenan, C. L. 1988. Long-term synaptic potentiation. *Science* 242:724–728.

Bullock, T. H., Orkand, R., and Grinnell, A. 1977. *Introduction to Nervous Systems.* San Francisco: W. H. Freeman and Company.

Dayhoff, J. E. 1987. Detection of favored patterns in the temporal structure of nerve cell connections. *IEEE First Int'l Conference on Neural Networks, Vol. III.*:63–77.

Dayhoff, J. E. and Gerstein, G. L. 1983. Detection of favored patterns in nerve spike trains. I. Detection II. Application. *J. Neurophysiol.* 49(6):1334–1363.

Dowling, J. E. 1987. *The Retina.* Cambridge, Massachusetts: Harvard University Press.

Gerstein, G. L., Perkel, D. H., and Dayhoff, J. E. 1985. Cooperative firing activity in simultaneously recorded populations of neurons: Detection and measurement. *J. Neuroscience* 5(4):881–889.

Hofer, M. A. 1981. *The Roots of Human Behavior.* San Francisco: W. H. Freeman.

Kistler, J., Stroud, R., Klymkowsky, M., Lalancette, R., and Fairclough, R. 1982. Structure and function of an acetylcholine recepter. *Biophys. J.* 37:371–383.

Lindsey, B. G. and Gerstein, G. L. 1979. Interactions among an ensemble of chorodotonal organ receptors and motor neurons of the crayfish claw. *J. Neurophysiology* 42(2):383–399.

Lindsey, B. G. and Gerstein, G. L. 1979. Proprioceptive fields of crayfish claw motor neurons. *J. Neurophysiology* 42(2):268–282.

Lindsey, B. G., Shannon, R., and Gerstein, G. L. 1989. Gravitational representation of simultaneously recorded brainstem respiratory neuron spike trains. *Brain Research* 483:373–378.

Lund, R. D. 1979. Tissue transplantation: A useful tool in mammalian neuroembryology. *Trends Neurosci* 3:xii–xiii.

MacGregor, R. J. 1987. *Neural and brain modeling.* San Diego: Academic Press.

Makowski, L., Casper, D. L. D., Phillips, W. C. and Goodenough, D. A. 1977. Gap junction structure II. Analysis of the x-ray diffraction data. *J. Cell Biology* 74:629–645.

Millhorn, D. E. and Hokfelt, T. 1988. Chemical messengers and their coexistence in individual neurons. *News in Physiol. Sciences* 3:1–5.

Parker, D. 1987. Optimal algorithms for adaptive networks: Second order back propagation, second order direct propagation, and second order hebbian learning. *Proc. IEEE International Conference on Neural Networks, Vol. II,* pp. 593–600.

Perkel, D. H., and Perkel, D. J. 1985. Dendritic spines: Role of active membrane in modulating synaptic efficacy. *Brain Research* 325:331–335.

Rall, W. 1978. Dendritic spines and synaptic potency. In *Studies in neurophysiology,* ed. R. Porter, pp. 203–209. Cambridge: Cambridge University Press.

Richmond, Optican, Podell, Spitzer. 1987. Temporal encoding. *J. Neurophys.* Jan. 57(1):132–145.

Rumelhart, D. E., Hinton, G. E., and Williams, R. J. 1986. Learning internal representations by error propagation. In *Parallel Distributed Processing,* Rumelhart, McClelland and the PDP Group, Chapter 8, pp. 318–362. Cambridge, Massachusetts: MIT Press.

Shepherd, G. M. 1988. *Neurobiology.* New York: Oxford University Press.

Spiral, Yarom, and Parnas. 1976. Modulation of spike frequency by regions of special axonal geometry and by synaptic inputs. *J. Neurophysiol.* 39:883–899.

Steiger, U. 1967. Uber den feinbau des neuropils im corpus pedunculatum der Waldameise. Zeitschrift für Zellforschung 81:511–536.

Thompson, R. F. 1985. *The Brain.* New York: W. H. Freeman.

9

The Kohonen Feature Map

Kohonen's self-organizing feature map is a two-layered network that can organize a topological map from a random starting point. The resulting map shows the natural relationships among the patterns that are given to the network. The network combines an input layer with a competitive layer of processing units, and is trained by unsupervised learning. This paradigm was presented by Kohonen, although seeds of the same idea appear elsewhere (Grossberg 1988, 1989). The examples and equations given in this chapter follow from Kohonen's work (1988).

Topological mappings of sensory and motor phenomena exist on the surface of the brain (as seen in Chapter 7). Kohonen's feature map can lend some insight into how a topological mapping can be organized by a neural network model. It is important to keep in mind, however, that brain mechanisms are different from the paradigm described here. The detailed structure of the brain is different, and input patterns are represented differently in biological systems. Furthermore, biological neural systems have a much more complex interconnection topology. However, the basic idea of having a neural network organize a topological map is illustrated effectively with the Kohonen feature map.

The Kohonen feature map finds the organization of relationships among patterns. Incoming patterns are classified by the units that they activate in the competitive layer. Similarities among patterns are mapped into closeness relationships on the competitive layer grid. After training is complete, pattern relationships and groupings are observed from the competitive layer. The Kohonen network provides advantages over classical pattern-recognition techniques because it utilizes the parallel architecture of a neural network and provides a graphical organization of pattern relationships.

BASIC STRUCTURE

The Kohonen feature map is a two-layered network. The first layer of the network is the input layer. Typically the second—competitive—layer is organized as a two-dimensional grid. All interconnections go from the first layer to the second; the two layers are fully interconnected, as each input unit is connected to all of the units in the competitive layer. Figure 9-1 shows this basic network structure.

When an input pattern is presented, each unit in the first layer takes on the value of the corresponding entry in the input pattern. The second layer units then sum their inputs and compete to find a single winning unit. The overall operation of the Kohonen network is similar to the competitive learning paradigm (described in Chapter 6), however, the Kohonen network differs in the details of its equations and the choice of weights to update on each training cycle.

Each interconnection in the Kohonen feature map has an associated weight value. The initial state of the network has randomized values for the weights. Typically the initial weight values are set by adding a small random number to the average value for the entries in the input patterns. (A graphical picture of this initial state will be described in the self-organization example.) The weight values are updated during the training of the network.

Figure 9-2 illustrates an example training set for the network in Figure 9-1. In this example, each input pattern is generated by a random process that makes each entry in the pattern vector uniformly distributed between 0 and 1, and the input pattern is a vector with n entries. As a result, the input patterns are uniformly spread over an n-dimensional hypercube. If $n = 2$, then the input patterns are uniformly spread over a square; such two-dimensional

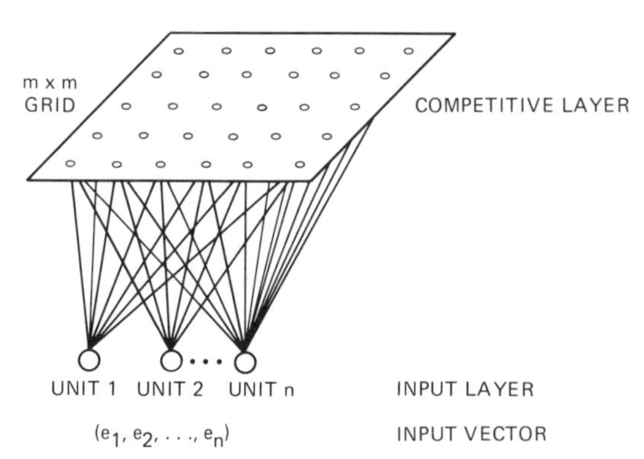

Figure 9-1. The basic network structure for the Kohonen feature map.

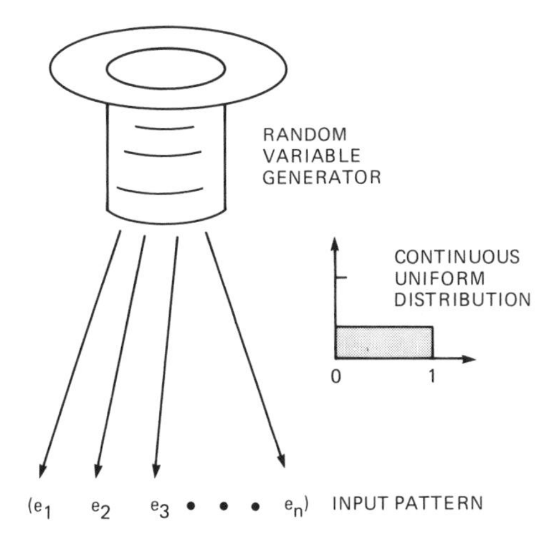

EXAMPLE INPUT PATTERNS:

(0.12, 0.71, 0.32, . . ., 0.56)

(0.24, 0.89, 0.91, . . ., 0.37)

(0.12, 0.90, 0.08, . . ., 0.49)

Figure 9-2. The random variable generator is depicted at the top as a hat from which numbers are drawn; these numbers are uniformly distributed between the values of 0 and 1. This random variable generator is used to provide numbers for each entry in each input vector.

input patterns will be used in our first example. In other examples and in actual applications, any set of patterns (with any value of n) may be used as inputs, and the patterns do not have to be uniformly spread.

An input pattern to the Kohonen feature map is denoted here as

$$\mathbf{E} = [e_1, e_2, e_3, \ . \ . \ . \ , e_n]$$

The connections from this input to a single unit in the competitive layer are shown in Figure 9-3. The weights are given by

$$\mathbf{U}_i = [u_{i1}, u_{i2}, \ . \ . \ . \ , u_{in}]$$

where i identifies the unit in the competitive layer. (These weights go to unit i. We identify the unit in the competitive layer by a single index, even though there is a two-dimensional grid of units in this layer.)

The first step in the operation of a Kohonen network is to compute a matching value for each unit in the competitive layer. This value measures the

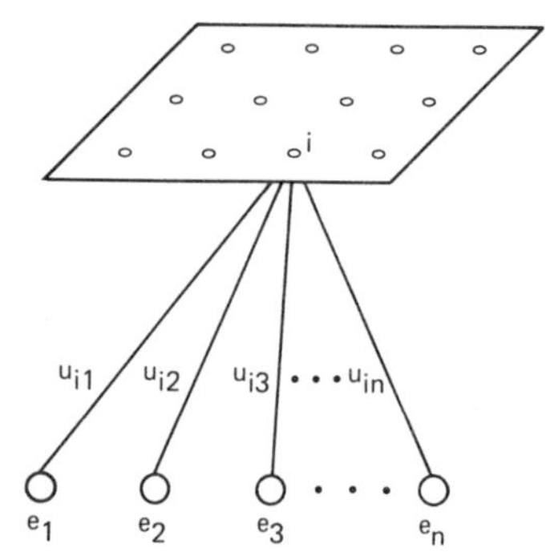

Figure 9-3. Connections from the input vector to a single unit in the competitive layer.

extent to which the weights of each unit match the corresponding values of the input pattern. The matching value for unit i is

$$\|\mathbf{E} - \mathbf{U}_i\| \tag{9-1}$$

which is the distance between vectors **E** and $\mathbf{U}_i$ and is computed by:

$$\sqrt{\sum_j (e_j - u_{ij})^2}$$

The unit with the lowest matching value (the best match) wins the competition. Here we denote the unit with the best match as unit c, and c is chosen such that

$$\|\mathbf{E} - \mathbf{U}_c\| = \min_i\{\|\mathbf{E} - \mathbf{U}_i\|\}$$

where the minimum is taken over all units i in the competitive layer. If two units have the same matching value from (9-1), then, by convention, the unit with the lower index value i is chosen.

After the winning unit is identified, the next step is to identify the neighborhood around it. The neighborhood, illustrated in Figure 9-4, consists of those processing units that are close to the winner in the competitive layer grid. The neighborhood in this case consists of the units that are within a square that is centered on the winning unit c. The size of the neighborhood changes, as shown by squares of different sizes in the figure. The neighborhood is denoted by the set of units N_c. Weights are updated for all neurons that

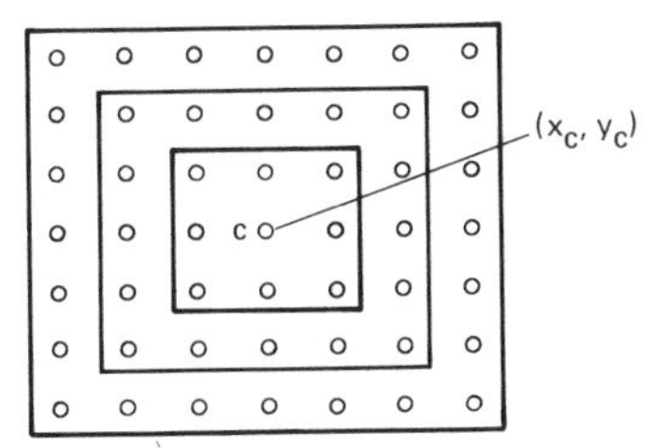

Figure 9-4. Neighborhood N_c, centered on unit c. Three different neighborhoods are shown: $d = 1, 2$, and 3.

are in the neighborhood of the winning unit. The update equation is:

$$\Delta u_{ij} = \begin{cases} \alpha(e_j - u_{ij}) & \text{if unit } i \text{ is in the neighborhood } N_c \\ 0 & \text{otherwise} \end{cases}$$

and

$$u_{ij}^{\text{new}} = u_{ij}^{\text{old}} + \Delta u_{ij}$$

This adjustment results in the winning unit and its neighbors having their weights modified, becoming more like the input pattern. The winner then becomes more likely to win the competition should the same or a similar input pattern be presented subsequently.

Note that there are two parameters that must be specified: the value of α, the learning rate parameter in the weight-adjustment equation, and the size of the neighborhood N_c.

The learning rate, α, begins initially at a relatively large value. During the learning process, α is decreased over a span of many iterations. The initial value of α is set by choice, and is denoted α_0. Typical choices are in the range $0.2-0.5$. The value of α is then decreased as training iterations proceed. An acceptable rate of decrease for α is specified by:

$$\alpha_t = \alpha_0\left(1 - \frac{t}{T}\right)$$

where t = the current training iteration and T = the total number of training iterations to be done. Thus α begins at a value α_0 and is decreased until it reaches the value of 0. The decrease is linear with the number of training iterations completed.

The size of the neighborhood is the second parameter to be specified. Typically the initial neighborhood width is relatively large, and the width is decreased over many training iterations. For illustration, consider the neighborhood in Figure 9-4, which is centered on the winning unit c, at position (x_c, y_c). Let d be the distance from c to the edge of the neighborhood. The neighborhood is then all (x,y) such that

$$c - d < x < c + d$$

and

$$c - d < y < c + d$$

This defines a square neighborhood about c. Sometimes this calculated neighborhood goes outside the grid of units in the competitive layer; in this case the actual neighborhood is cut off at the edge of the grid.

Since the width of the neighborhood decreases over the training iterations, the value of d decreases. Initially d is set at a chosen value denoted by d_0. Typical values for d_0 may be chosen at a half or a third of the width of the competitive layer of processing units. The value of d is then made to decrease according to the equation

$$d = \left[d_0 \left(1 - \frac{t}{T} \right) \right]$$

where $t =$ the current training iteration and $T =$ the total number of training iterations to be done. This process assures a gradual linear decrease in d, starting with d_0 and going down to 1. The same amount of time is spent at each value.

In summary, the basic rules of the Kohonen feature map can be described qualitatively as follows:

1. Locate the unit in the competitive layer whose weights best match the input pattern.
2. Increase matching at this unit and its neighbors by adjusting their weights.
3. Gradually decrease the size of the neighborhood and the amount of change to the weights as the learning iterations progress.

EXAMPLES

A Self-Organization Example

Figure 9-5a shows the network used to self-organize a two-dimensional map that reflects the distribution of input patterns. The input layer has two processing units, and the competitive layer is an 8×8 grid of units.

The training patterns are vectors with two entries. Each entry ranges from 0 to 1 and is chosen from a uniform distribution for each new training pattern (as in Figure 9-2). As a result, each training pattern is a randomly selected position on the square plot shown in Figure 9-5b.

The initial weights of the network are set to the value 0.5 plus a small—within 10%—randomized value. Figure 9-6a shows a plot of these initial weights. Each unit in the competitive layer is a point on this graph. The

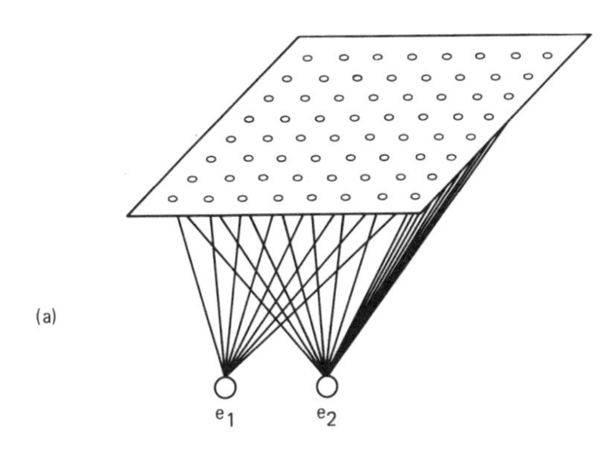

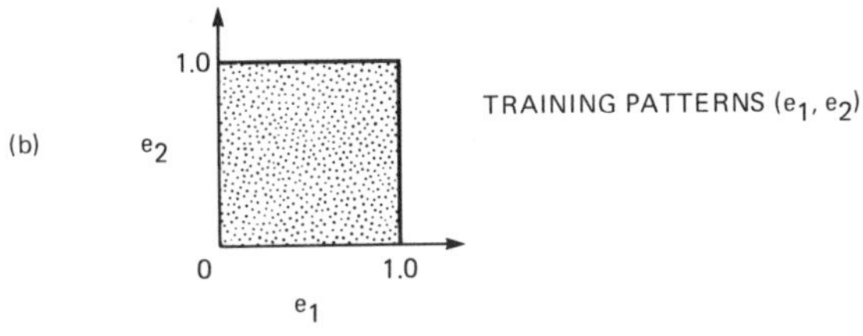

Figure 9-5. (a) Network used to self-organize a two-dimensional map that reflects the distribution of input patterns. (b) Training patterns, uniformly distributed on the square.

coordinate values of this point are the values of the incoming weights for the unit. Thus, (w_{i1}, w_{i2}) is plotted for each competitive unit i.

Figure 9-6a also shows two of the plotted points connected together. These points corrspond to two adjacent units in the competitive layer, shown connected in Figure 9-6b. In later plots, all pairs of units in the competitive layer that are adjacent will be connected. This will allow us to see how the pattern of weights changes as the network organization evolves during training.

The Kohonen network gradually organizes, starting from the initial position in Figure 9-7a. The cluster of points in the center of the graph of Figure 9-7a depicts the randomized initial weight values. Figure 9-7b shows the network after 1,000 iterations. Here, the natural ordering among the units in the competitive layer has been found. Although the points plotted are based on the connection weights, the units are interconnected in a grid. This interconnection pattern is a compressed and distorted rendition of the original grid of

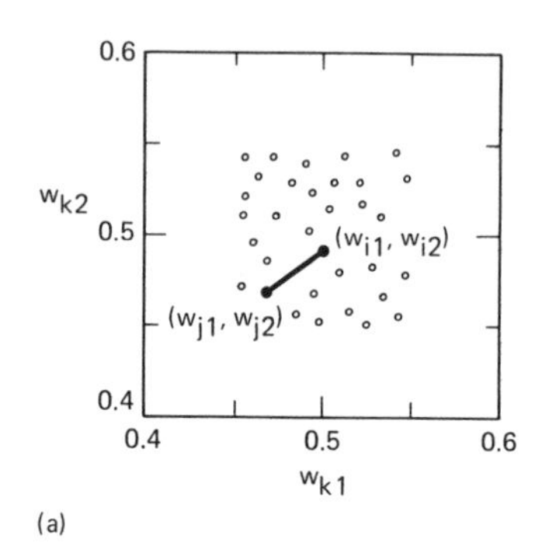

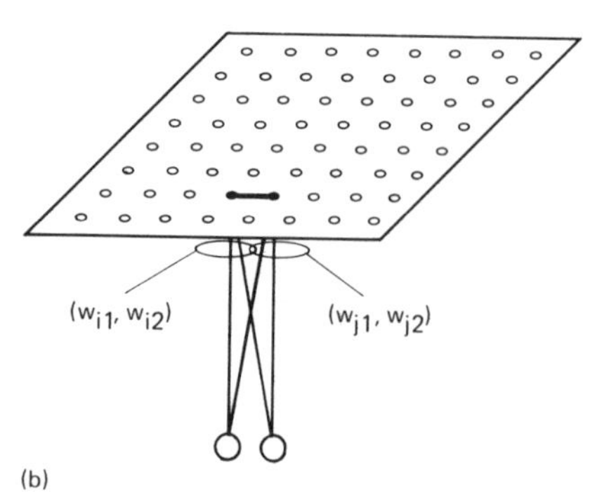

Figure 9-6. (a) Initial weights for the network in Figure 9-5a. (b) Adjacent units in the competitive layer are connected in the plot of Figure 9-6a.

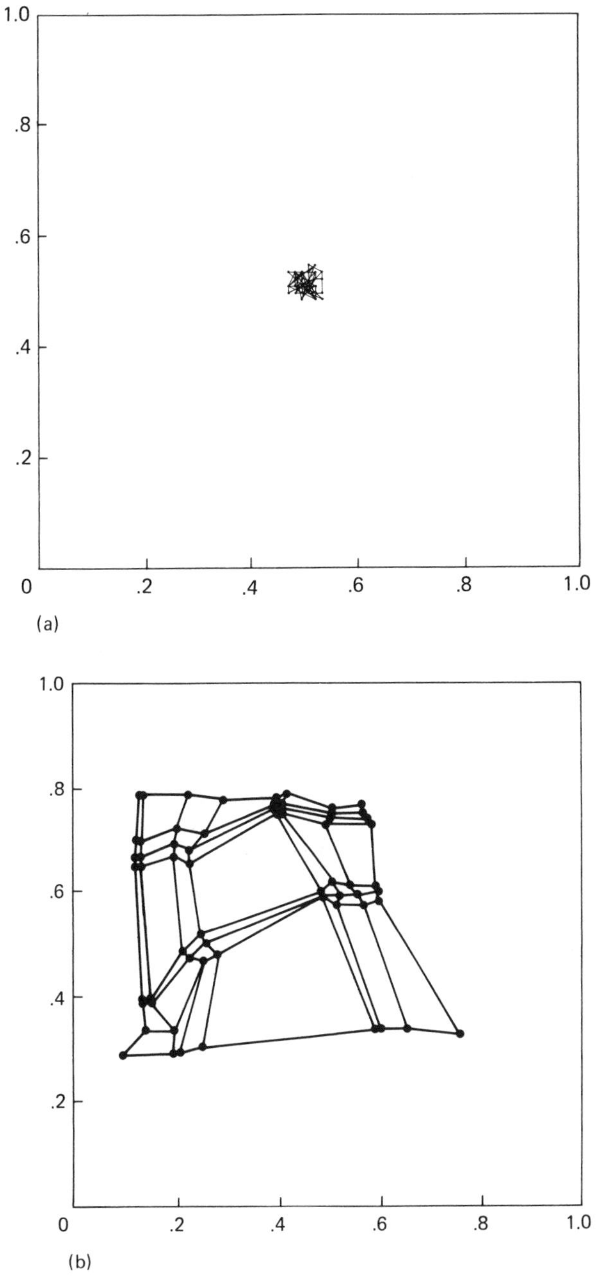

Figure 9-7. (a) Initial position for weight vectors of the network in Figure 9-5a. (b) Weight vectors after 1,000 training iterations.

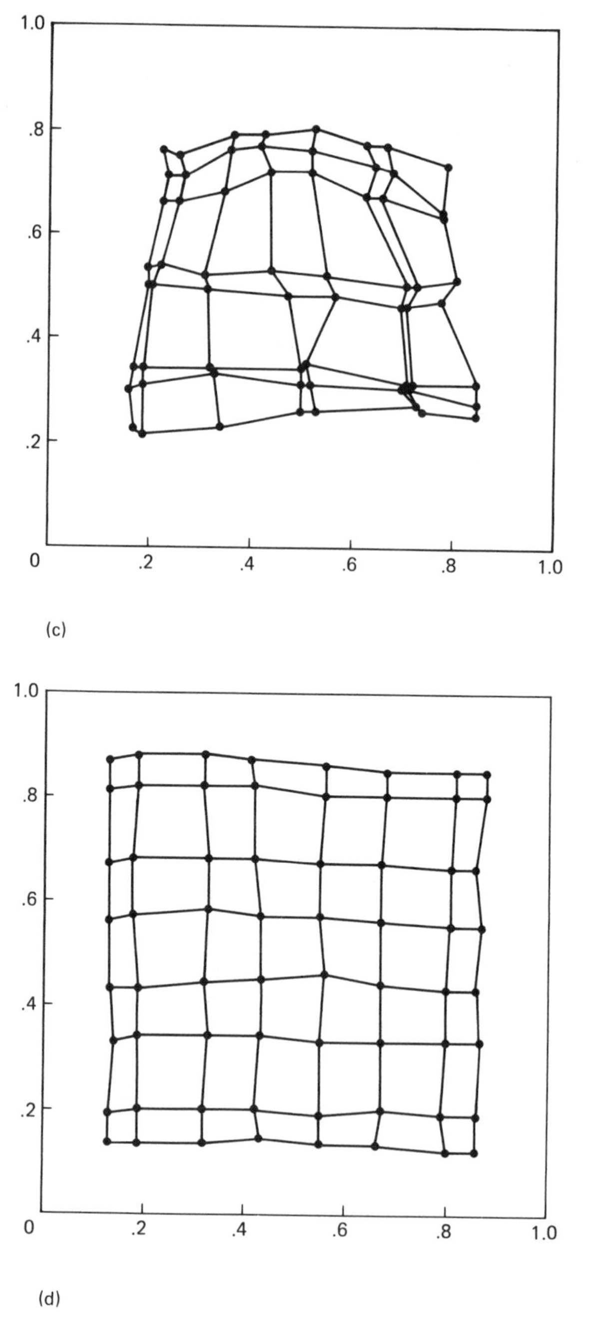

Figure 9-7. (cont.) (c) Weight vectors after 6,000 training iterations. (d) Final weight vectors, after 20,000 iterations. This network was organized with parameters $\alpha_0 = 0.2$ and $d_0 = 4$.

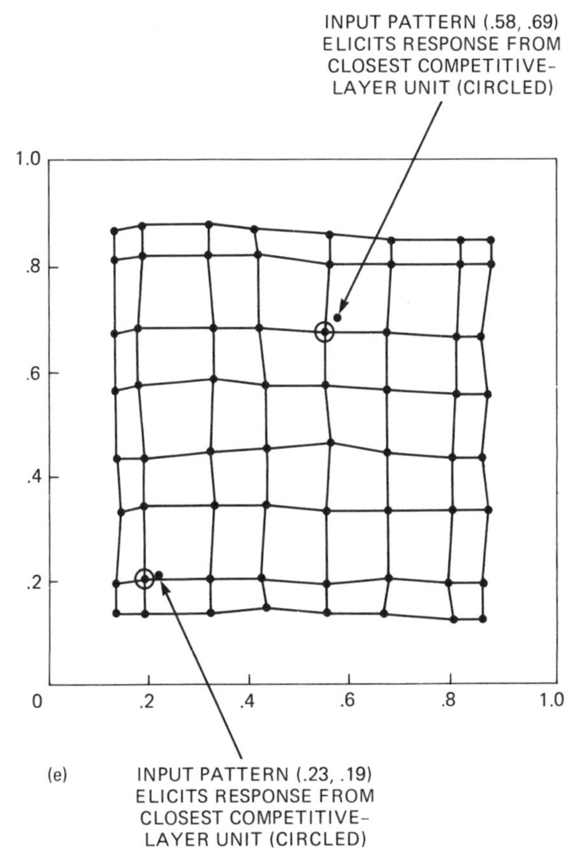

Figure 9-7. (cont.) (e) Response of the trained network to two different input patterns. The winning unit is closest to the input pattern on the graph. Two input patterns are shown as points, with the nearby winning unit circled.

units in the competitive layer. (Why the network should organize its weights in this fashion is discussed in the One-Dimensional Example section.)

Figure 9-7c shows the state of the network after 6,000 iterations. The unit weights have spread out more over the total space available and retain the same ordering that they found in the prior graph. The spacing of the points is not uniform yet, however. Figure 9-7d shows the final state of the network—after 20,000 iterations. The units still have the natural ordering found earlier, but have spread out to cover most of the square uniformly. Each axis of the square in Figure 9-7d goes from 0 to 1 because this is the range of the entries in the input patterns. To interpret these results further, Figure 9-7e shows the response of the trained network to example input patterns. When an input pattern is presented to the network, the competitive unit that wins is the closest unit on the graph in Figure 9-7e. Thus Figure 9-7d illustrates the response of the network to the possible input patterns. The network has developed a maplike representation in which there is a mapping from each input pattern to a unit in the competitive layer grid (the closest unit in Figure 9-7e). Furthermore, two input patterns that are similar will activate two units in the competitive grid that are close in Figure 9-7d.

Note that Figure 9-7d shows a slight edge effect—the boundaries of the grid are some distance from the edges of the square. As a result, the outer border is slightly compressed. Because no input vectors occur outside the edge of the graph, there is less influence pulling the outer ring outward. Thus the outer ring of points appears to "pull away" from the edges.

Parameters for the runs in Figure 9-7 were $\alpha_0 = 0.2$ with d set initially to 4 and decreasing to reach 1. A total of 20,000 iterations were done, and each iteration consisted of the presentation of a single training pattern.

The type of organization shown in Figure 9-7d is found quite reliably. The order of the units in the competitive-layer grid may be inverted in the graph, but the usual result is an orderly array as shown. Occasionally, however, opposite sides can move to an ordering that is in the opposite direction. The resulting map (Figure 9-8) is not as effective because gaps appear on the graph near the crossover point; no units will be a close match to input patterns that occur in these gaps. The organization of this faulty map depends partly on initial state of the weights and partly on the sequence of training patterns.

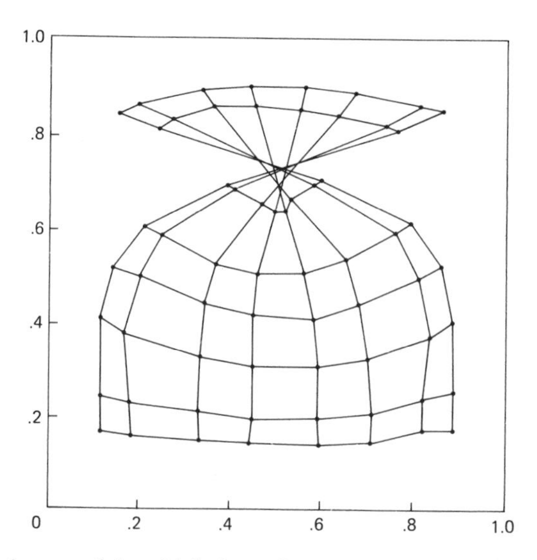

Figure 9-8. A network in which the ordering crosses over. Weights were randomized initially, and 20,000 iterations were done with $\alpha_0 = 0.2$ and $d_0 = 3$.

Self-organization of the Kohonen feature maps appears to be done in two stages: first the network finds its initial ordering and then it spreads its competitive-layer units so that each responds to approximately the same number of input patterns. The second stage usually takes at least 10 times the number of iterations required for the first. Figure 9-7b shows the result after the first stage is completed, where the basic ordering among points has been established; Figure 9-7d shows the results after the second stage is completed and the points are spread out evenly.

Mapping Two Dimensions Into One

A Kohonen feature map can be organized to allow a one-dimensional chain of units to span a pattern space in two dimensions. This example illustrates an important property of Kohonen networks — the ability to reduce the number of dimensions required to encode a particular set of input patterns. Figure 9-9a shows a network with two input units and a linear chain of 40 competitive output units. The two input units allow for two-dimensional patterns to be input to the network, and the network maps these patterns to the one-dimensional chain of output units.

Training patterns were chosen from a uniform distribution in which each entry was between 0 and 1 (as in Figure 9-2). A total of 60,000 pattern presentations were done to achieve the final state. Figures 9-9b–9d show different stages of organization for this network, with the final trained network presented in Figure 9d. The chain begins in a twisted state, with many crossovers (Figure 9-9b); this state is from the initial random values for the weights. The next stage of training results in an ordering that allows the chain to unfold from its inital twisted position (Figure 9-9c). Subsequently the chain spreads across and around the square to cover the total area (Figure 9-9d).

Nonuniform Probability Densities

If the input patterns have an uneven probability density, then the competitive layer of processing units will reflect the distribution of those patterns after adaptation is completed. The result is that the ability of the competitive layer of units to discriminate between similar patterns will be greater for those patterns that occur more frequently. Thus the network can distinguish finer differences among patterns that occur more often. Figure 9-10 shows a specific example of this trend, based on a network that has two input units and an 8×8 grid of output units (Figure 9-5a).

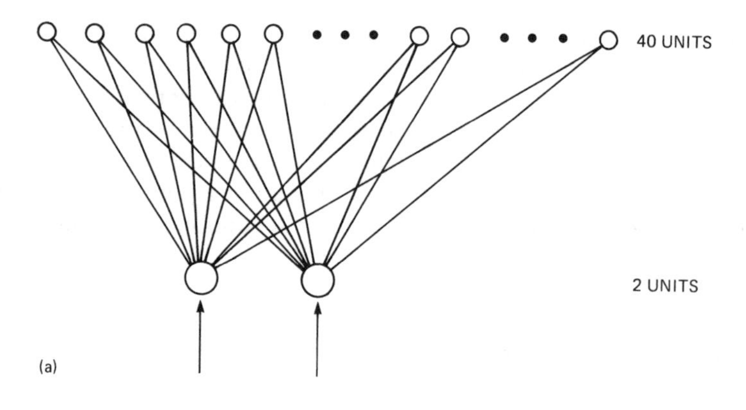

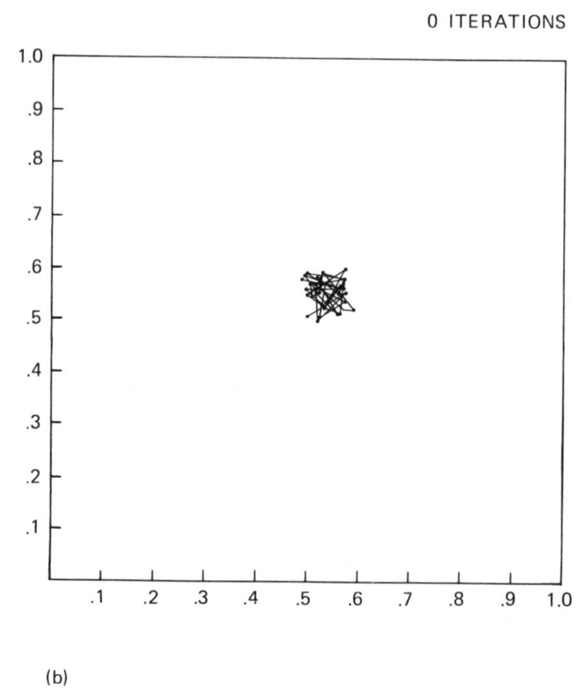

Figure 9-9. (a) Network with a two-dimensional input vector and a chain of 40 units in the competitive layer. (b) Initial state of the network with random weight values.

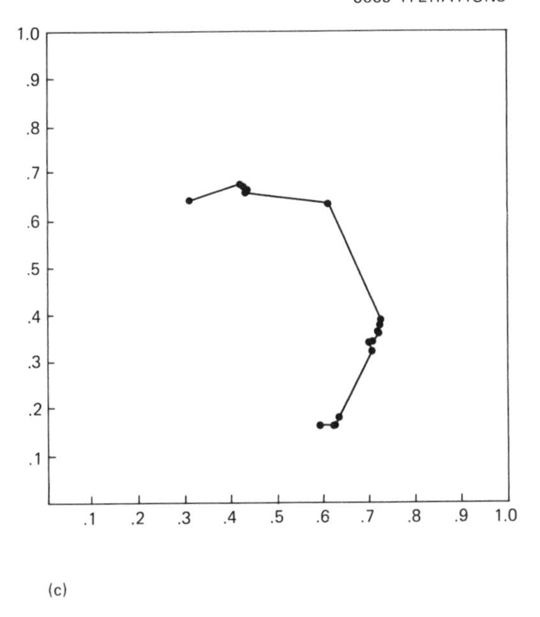

(c)

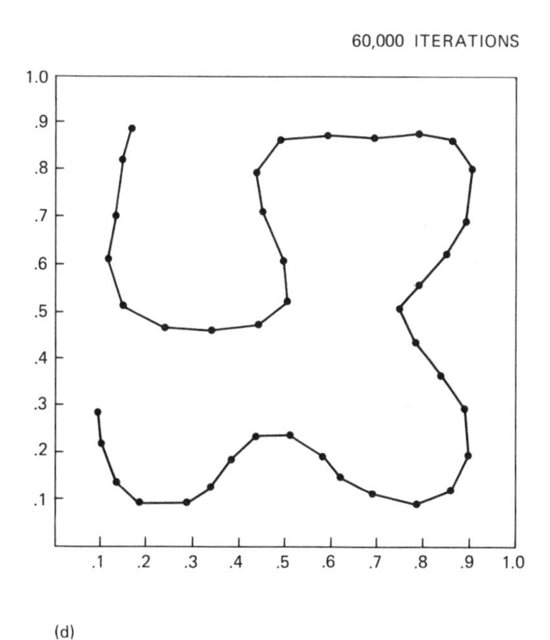

(d)

Figure 9-9. (cont.) (c) Intermediate stage in network organization. Each point graphs the incoming weights of an individual top-layer unit; the points are connected in the corresponding chain. (d) Final state after 60,000 iterations, using $\alpha_0 = 0.2$ and $d_0 = 14$.

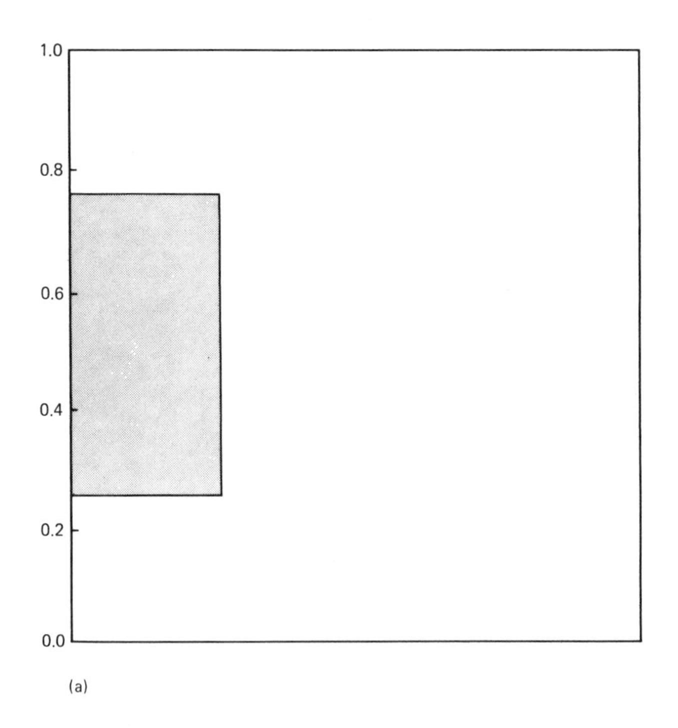

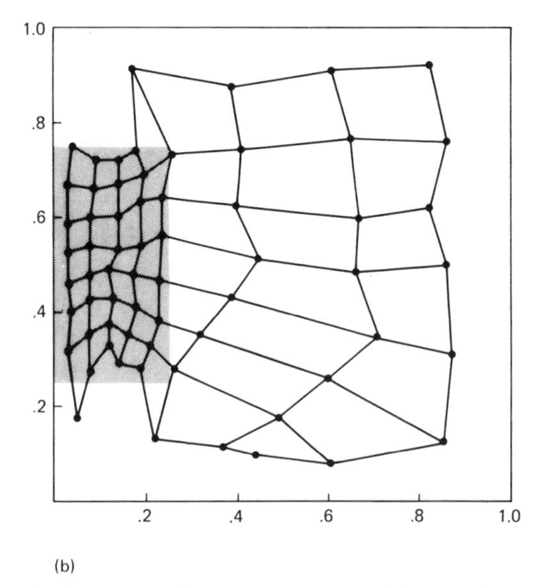

Figure 9-10. (a) Structure of input patterns used for training a network with input units and an 8 × 8 grid of output units. The shaded region had 42% of the input pattern density, uniformly spread across the area. The remaining area had 58% of the patterns, uniformly spread over the area. (b) Result after 20,000 iterations of training. Parameters used were $\alpha_0 = 0.2$ and $d_0 = 2$.

Figure 9-10a illustrates the structure of the input patterns used. The density of patterns is greater in the shaded region at the left—42% of the input patterns were from this area; the remaining input patterns were from the nonshaded area. The patterns in each area were evenly distributed across their respective areas.

Figure 9-10b shows the same structure after 20,000 iterations of training. The regular crosshatch (seen in Figure 9-7d) has been distorted, with more of the output units appearing in the area where patterns occur more frequently. As a result, the competitive units discern finer differences among the patterns in the shaded area—the area where more patterns occur. An input pattern, when presented to the trained network, will activate the output unit that is closest to that input pattern. Suppose the input pattern appears in the shaded area when graphed. Then, the output unit that corresponds to the point that is closest will win the competition. Since output units are spaced more closely in the shaded area, they will distinguish finer differences among input patterns in that way.

Mapping Different Dimensions

The most interesting uses of the Kohonen feature map occur when mapping patterns from one dimension to another. This type of transformation is important because it can be used to reduce the dimensionality of the data. Further interest arises from the unusual graphical contours that are sometimes produced by the trained networks.

In the Kohonen feature map, the dimension of the input pattern is the number of entries in the input pattern vector. The dimension of the output is the number of dimensions to the grid (a line, plane, three-dimensional array, etc.) of competitive units. Figure 9-7 shows a network that has the same dimension for both the inputs and the outputs. Figure 9-9 shows a network that has a different dimension for the inputs and the outputs—patterns in two dimensions are mapped into a one-dimensional chain.

Figure 9-11 illustrates a Kohonen feature map that maps three dimensions into two, thus reducing the dimensionality of the patterns. There are three input units so that a three-dimensional data vector can be input. The output units are organized into a two-dimensional grid. Training the network resulted in a grid that appeared as though it were a two-dimensional piece of rubber that was pushed inward to form ruffles that span a three-dimensional volume. The units on this grid are spaced in an attempt to span the three dimensions evenly. From an applications standpoint, this is an interesting result because the three-dimensional patterns can be classified by units that respond to subsets of those patterns that are relatively evenly spaced and that

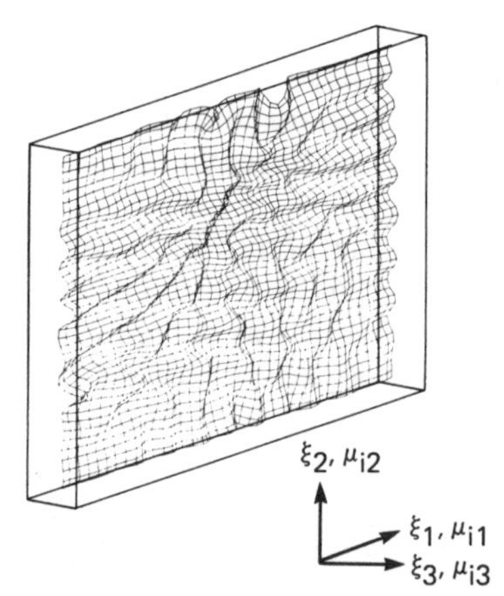

Figure 9-11. Weight vectors for a trained Kohonen feature map that maps three dimensional input into a two-dimensional grid of top-layered units. (From T. Kohonen. *Self-Organization and Associative Memory.* Springer-Verlag 1988.)

reflect the structure of the input patterns. From a biological standpoint this is an interesting result because it is reminiscent of certain brain maps that appear to map multidimensional data into the two-dimensional surface of the cerebral cortex (Kohonen 1988b).

A One-Dimensional Example

So far we have shown various examples of the Kohonen feature map, but have not yet attempted to show why this paradigm works the way it does when it finds a natural ordering and classification of input patterns. Relevant proofs appear elsewhere (Cottrell and Fort 1987; Orey 1971; Ritter and Schulten 1986), and are too lengthy to include here. Instead we present an intuitive argument that gives the basic idea of how the Kohonen feature map produces its organization.

We start with a simple example network that maps a one-dimensional "pattern" into a linear row of output units. First we will show how the feature map starts with a random ordering and proceeds to eliminate "kinks" or bends by its adaptive process using a network with one input unit and eight competitive units (see Figure 9-12a). Initially, the weights are randomized and the adaptive process slowly orders these weights so that they are in ascending or descending order. Figure 9-12b shows an initial state, with weights to each output unit graphed. A final state, in ascending order, is shown in Figure 9-12c.

The input to the network is denoted by (e). This input pattern consists simply of a vector with a single entry. Each pattern is generated randomly from the interval [0,1]. The best match is unit c, with c chosen such that:

$$u_c - e = \min_j |u_j - e|$$

where the minimum is taken over all units j in the competitive layer. Then c is the winning unit.

The neighborhood N_c consists of all units in the following interval:

$$[c - d, c + d]$$

Units in this interval have their weights updated by

$$u_j^{\text{new}} = u_j^{\text{old}} + \alpha(e - u_j)$$

where α is the learning rate parameter. The adaptive process gradually has the weights assume new values so that they become ordered. The resulting order is a sequence of weights that is either ascending or descending.

To illustrate how this ordering comes about, we will look at a specific part of the chain of output units consisting of a five unit sequence. To avoid any effects due to edges, we will assume that this subset of the output units is not at either boundary of the output layer. There are four pairs of adjacent units, each with two possible orderings: A unit's weight can be higher or lower than the previous unit's weight (equality is unlikely). Thus there are a total of 2^4, or 16 possible orderings for the sequence of five units. These 16 possibilities are shown in Table 9-1.

We will analyze 3 of these 16 cases to illustrate how ordering is established. The illustration provides the main concepts behind the ordering process but is not rigorous. A more rigorous analysis would involve analysis of all 16 cases.

First we define a measurement that reflects the degree of disorder in the output unit weights. This measurement is calculated by

$$D = \sum_{i=2}^{t} |u_i - u_{i-1}| - |u_t - u_1|$$

where the absolute values of differences of successive weights are added and compared to the difference between the first and the last weights. Thus a sequence of weights that alternately goes up and down will result in a higher sum than a sequence of weights that is either increasing or decreasing. The higher sum yields a higher value of D, reflecting the disorder.

TABLE 9-1. ALL POSSIBLE RELATIONSHIPS BETWEEN THE WEIGHTS OF FIVE SUCCESSIVE UNITS.

Case	u_{c-2} ? u_{c-1}	u_{c-1} ? u_c	u_c ? u_{c+1}	u_{c+1} ? u_{c+2}
1	>	>	>	>
2	>	>	>	<
3	>	>	<	>
4	>	>	<	<
5	>	<	>	>
6	>	<	>	<
7	>	<	<	>
8	>	<	<	<
9	<	>	>	>
10	<	>	>	<
11	<	>	<	>
12	<	>	<	<
13	<	<	>	>
14	<	<	>	<
15	<	<	<	>
16	<	<	<	<

The value of D is always greater than or equal to 0 ($D \geq 0$). Furthermore, D is 0 exactly when $u_1, u_2, \ldots, u_t$ form an ascending or descending sequence. The degree of disorder D gets lower when the sequence becomes more ordered. The key to the ordering proof depends on the fact that D more often decreases than increases during updating. This is true assuming that e is a random variable from the interval [0,1]. Over the course of many updating iterations, D gradually gets smaller, and eventually becomes equal to 0.

The illustration that follows uses a neighborhood parameter $d = 1$, which reflects a neighborhood of three units. As already noted, we analyze a chain of five units. Assume that the central unit of the five has been chosen as the competitive winner; its neighborhood is then three units. Each unit has its weight modified on each updating iteration. All five units, then, have weight differences between adjacent units affected on each updating.

The first case to be analyzed is shown in Figure 9-13a; this case is entry 11 of Table 9-1. The solid line in both graphs of the figure shows the values of the weights consistent with entry 11, with unit 1 $<$ unit 2 $>$ unit 3 $<$ unit 4 $>$ unit 5. Note that there are two peaks when the weight values of successive units are graphed for this case. The top and bottom graphs of Figure 9-13a show two different situations, depending upon whether the input signal e is larger or smaller than u_c. In the top graph, e is greater than the weight

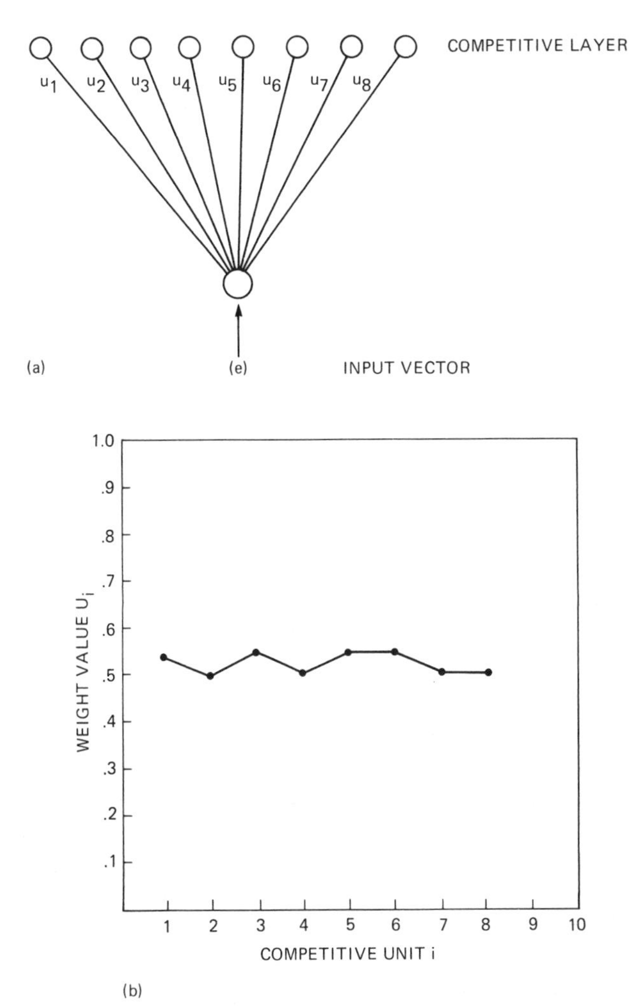

Figure 9-12. (a) A network with a single input unit and eight competitive units. (b) Initial state of successive weight values.

connected to the middle unit u_c, and in the bottom graph e is less than u_c. The open circles and dotted lines show the weight values after updating. In both situations the degree of disorder decreases and the sum D becomes smaller. The decrease in disorder can be observed geometrically from Figure 9-13a; both graphs are closer to having a descending sequence.

The second case to be analyzed, entry 13 in Table 9-1, is shown in Figure 9-13b. In this case the initial weights for the sequence of units shows a single

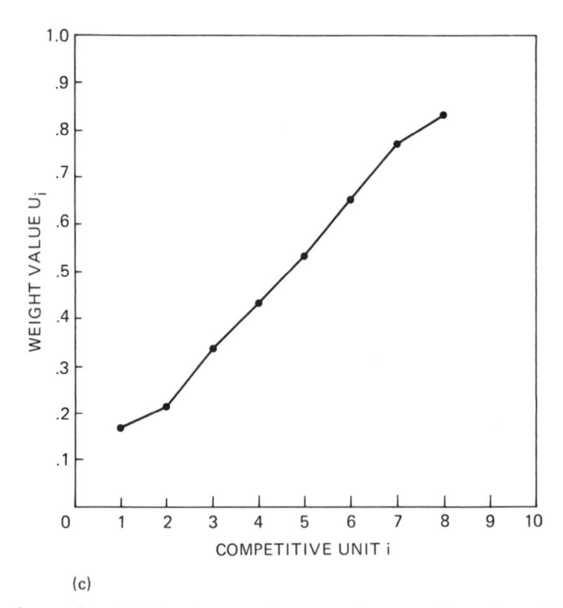

Figure 9-12. (cont.) (c) Final state of successive weight values. Five hundred iterations were done with $\alpha_0 = 0.3$ and $d_0 = 4$.

peak in the center, with weights ascending on the left and descending on the right. Two possible updating situations are again shown, depending on whether e is larger or smaller than u_c. When e is larger, the degree of disorder D increases (top graph). When e is smaller, the degree of disorder D decreases (bottom graph). These two results can be seen geometrically in the figure, as the peak becomes higher (and hence more disordered) at the top, and lower (less disordered) at the bottom.

The third case, shown in Figure 9-13c, is taken from entry 16 in Table 9-1. Here the initial weights form an ascending sequence. Updating produces no change in the degree of disorder; D is 0 before and after the update to the neighborhood N_c. This case is called an "absorbing state": If weights to the five units are ordered in an ascending or descending sequence, then the weights will stay ordered when the center unit u_c has its neighborhood updated. This case is key to the fact that once a network attains a natural ordering, the network keeps that natural order.

A detailed analysis of all the possible cases in Table 9-1 shows that for more than half of the cases D decreases. From a probability standpoint, therefore, a new input is more likely to make the network more orderly and hence lower D. On the average, the network moves toward a more ordered state as new inputs are presented and weights are updated. This movement is reflected by a

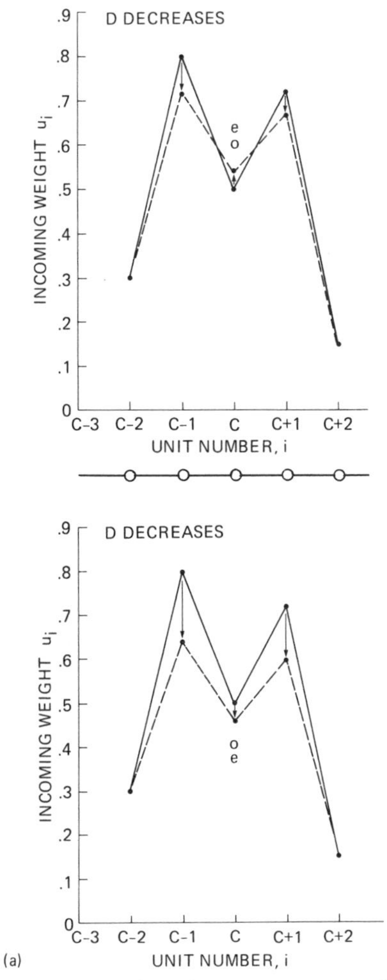

Figure 9-13. (a) A case in which D decreases upon updating. The solid line connects the original values of weights; the dotted line connects the updated values of weights.

decrease in D. Eventually the chain of units is either an increasing or decreasing sequence.

Sometimes a network reaches a point where part of its weights are in increasing order and part are in decreasing order (see Figure 9-13d). It is important to note that in this situation the network will still find a natural ordering even though a sequence of five increasing weights or five decreasing weights is an absorbing state. When the unit at the peak or next to the peak in

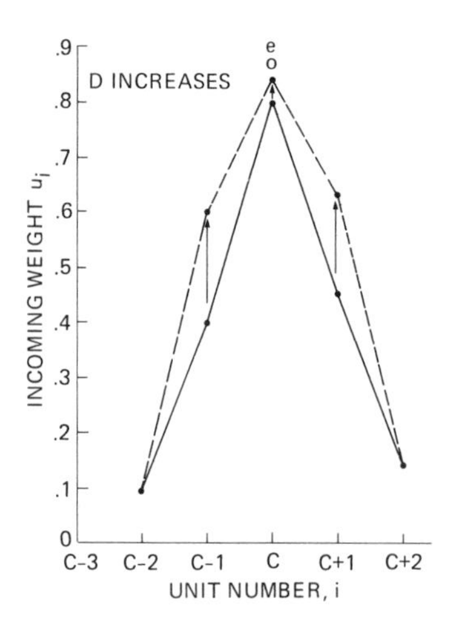

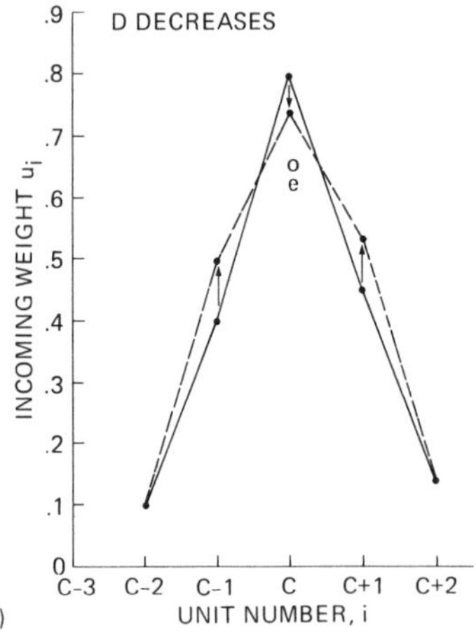

Figure 9-13. (cont.) (b) In this case, D increases if $e > u_c$, and decreases when $e < u_c$.

Figure 9-13d is the winner, the resulting update will cause the overall sequence to become less disordered. Thus, the chain of units will eventually become unbent, and the network will be able to reach a final ordered state.

We have not been rigorous here in our arguments, nor have we covered the general case. However, we have provided the basic idea for how a natural ordering is found by the network. A rigorous proof requires examining all 16 cases of Table 9-1, in addition to edge effects. A complete proof would require allowing for different dimensions in the input vector and the output grid, and for varying probability densities of input vectors.

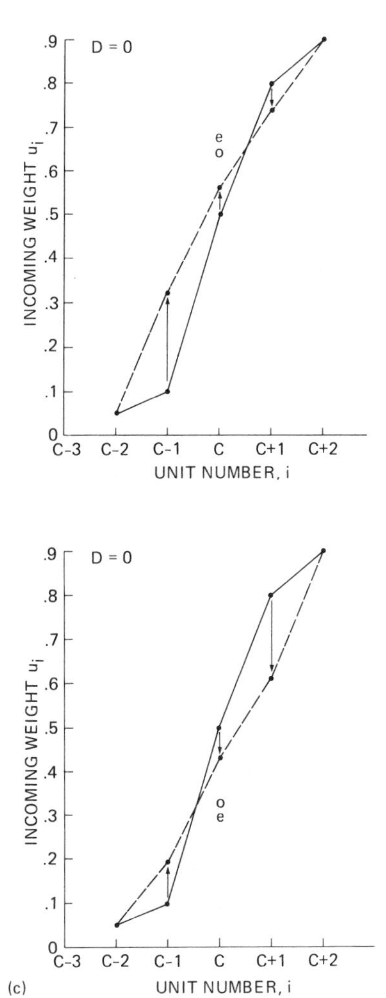

Figure 9-13. (cont.) (c) A case where D = 0 before and after updating. Here the sequence is ascending. Parameter $\alpha = 0.4$.

As mentioned before, there are two stages in the training of the Kohonen feature map. In the first state, the network finds its natural ordering of input patterns. We have illustrated the basic idea of how this ordering comes about. The second stage of training allows the competitive layer units to spread out until they span the pattern space and also reflect the relative distributions of input patterns.

The second stage can be visualized intuitively: The weights of the units in the winning neighborhood N_c move toward the input pattern as updating takes place. As shown in Figure 9-7, the competitive layer weights start clus-

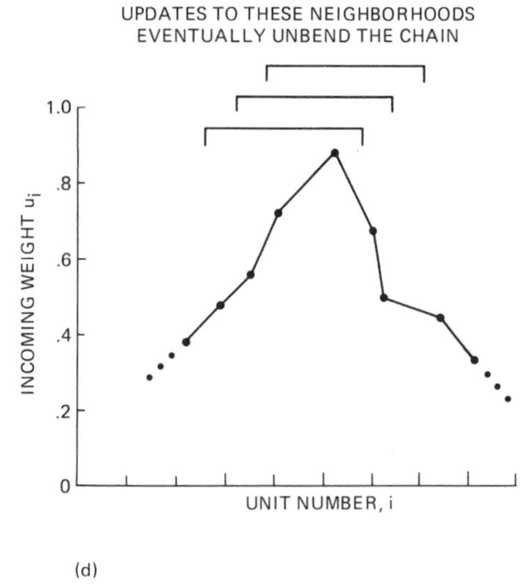

Figure 9-13. (cont.) (d) An ascending sequence followed by a descending sequence. Updates to the central neighborhoods eventually unbend the chain.

tered at the center then spread out. Since updating produces a movement toward the input patterns, the competitive layer units in the long run tend to reflect the densities of the different input patterns (as illustrated in Figure 9-10; Cottrell and Fort 1987; Kohonen 1988; Orey 1971; Ritter and Schulten 1986). When the input vectors are distributed uniformly, the competitive layer units tend to spread out uniformly to cover the available pattern space. When the input patterns have a nonuniform structure the output units tend to reflect that structure, even if the input patterns have a higher dimension than the output layer grid.

Note that there are two opposing tendencies in the Kohonen feature map: The weight vectors tend to describe the density function of the input vectors, and there is a local interaction between processing units. The local interactions occur because all units in a neighborhood are updated at the same time. The result is a tendency to preserve continuity in responses of competitive layer units to the sequences of input vectors.

An Applications Example

A striking use of the Kohonen feature map is found in the phoneme maps developed by Kohonen (1988a, 1989) built for Finnish and Japanese. In this application, the feature map is trained to distinguish different spoken phonemes, the basic units of speech sounds. Different units in the competitive layer respond to different phonemic inputs. Kohonen combines his neural network with appropriate preprocessing and postprocessing to produce a system that correctly recognizes more than 90% of the incoming sounds.

The network in this example has two layers, with 15 input units and 96 competitive units. It was trained on continuous speech that was sampled at time intervals of about 8 ms. The auditory input was sampled and then subjected to fast Fourier transform preprocessing, resulting in a 15-component vector that represented the signal energy in each of 15 frequency bands. The language used for training was Finnish, which has 21 distinct phonemes. The continuous speech records used for training included standard phonemes as well as the many transition sounds that tend to occur during normal continuous speech.

In the trained network the competitive layer units each respond best to a particular pattern input. Because the same phoneme can vary considerably in its acoustic characteristics, different units in the competitive layer will win the competition when slightly different forms of the same phoneme are presented. In addition, the same unit will sometimes win the competition for more than one phoneme. It is possible to map these phonemes across the competitive layer of the trained network. The resulting map becomes a "similarity map" in which the distance between two units is approximately proportional to the dissimilarity of sounds to which the different units respond. However, this is not a mathematically rigorous relationship; the network simply tends to produce this type of organization during training.

Figure 9-14 shows the response of the competitive layer to the Finnish word *humppila*. A word here is treated as a sequence of individual sounds. (The data is time-windowed.) This sequence of sounds then activates a sequence of units in the competitive layer. There is a smooth, connected activation pattern in the competitive layer because units that are near one another respond to similar patterns, and the input word has a relatively continuous transition from one sound to another. The data in Figure 9-14 were also smoothed by use of a moving average of the sequential locations.

Auxiliary maps are used for special sounds that are more difficult to distinguish. These sounds include transients such as certain consonant-to-vowel transitions, particularly stop consonants, which change much more rapidly than other phonemes and are strongly affected by adjacent vowels. In an auxiliary map, the entire map is dedicated to only a few different sounds, and its competitive units can make finer distinctions among those sounds. The auxiliary map thus provides higher resolution for particular phonemes. Context-sensitivity is handled with a separate postprocessing technique (Kohonen 1989), independent of auxiliary maps.

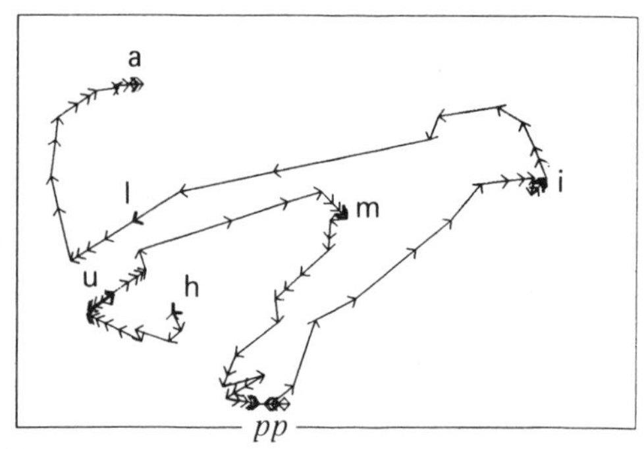

Figure 9-14. Trajectory of the response to the Finnish word ***humppila*** over the map. A moving average of the sequential locations was computed in order to make the curves smoother. (From Kohonen in *Neural Computing Architectures,* ed by I. Aleksander, MIT Press.)

The phoneme maps studied illustrate the role that a neural network can play in automated speech recognition. They show a tangible example of the use of the Kohonen feature map, and show how a segment of speech can result in an activation pattern in the map. Combined with preprocessing and post-processing that use conventional techniques, even a relatively simple neural network can provide the accuracy needed for a realistic application (Kohonen 1989).

OVERVIEW

The Kohonen feature map illustrates an important phenomenon that occurs in biology—the topographical map on a two-dimensional surface that represents sensory or motor events. In biological systems, the maps are folded irregularly across the convoluted surface of the cerebral cortex. These topographical maps occur for the visual and auditory systems, for motor control, and for other sensory systems. A pattern of activation across the sensory maps occurs in response to a time-varying input; organizations of biological maps are developed so that more space is dedicated to patterns that occur more often.

Artificial topographical maps are not the same as their biological counterparts. The input representations in biology are quite different from those of Kohonen feature maps, and biological neurons communicate via pulses that appear to contain information not passed in artificial neural connections. The details of synaptic adjustment and interneuronal competition are not known in biology, and may not be the same as that of artificial self-organizing maps.

The Kohonen feature map does, however, provide a neural network model for the adaptive organization of useful topological maps. In both biological and artificial systems, topological feature maps provide useful components for a complex pattern-recognition system.

References

ANSim User's Manual, Version 1.2, SAIC, 10260 Campus Point Drive, San Diego, California 92121.

Cottrell, M., and Fort, J.-D. 1987. *Ann. Inst. Henri Poincare* 23, 1.

Grossberg, S. 1987. *The Adaptive Brain* (Vols. I & II) New York: North-Holland.

Grossberg, S. 1988. *Neural Networks and Natural Intelligence.* Cambridge, MA: MIT Press.

Kohonen, T. 1988a, March. The neural phonetic typewrite. *Computer* 21(3): 11–22.

Kohonen, T. 1988b. *Self-Organization and Associative Memory* (2nd ed.). New York: Springer-Verlag.

Kohonen, T. 1989. Speech recognition based on topology-preserving neural maps. In *Neural Computing Architectures,* ed I. Aleksander, pp. 26–40. Cambridge, MA: MIT Press.

Orey, S. 1971. *Limit Theorems for Markov Chain Transition Probabilities.* London: Van Nostrand.

Ritter, H. and Schulten, K. 1986. *Biological Cybernetics* 54: 99.

Tattershall, G. 1989. Neural map applications. In *Neural Computing Architectures,* ed. I. Aleksander, pp. 41–73. Cambridge, MA: MIT Press.

10

Counterpropagation

Counterpropagation was originally proposed as a pattern-lookup system that takes advantage of the parallel architecture of neural networks (Hecht-Nielsen 1987). Counterpropagation is useful in pattern-mapping and pattern-completion applications and can also serve as a sort of bidirectional associative memory. Candidate applications include pattern classification, function approximation, statistical analysis, and data compression.

Counterpropagation nets are usually trained to perform pattern mapping, the mapping of one pattern to another for an entire set of patterns. When presented with a pattern, the trained network classifies that pattern into a particular group by using a stored reference vector; the target pattern associated with the reference vector is then output. The activity of the hidden layer of units is key in this paradigm, as the hidden layer performs a competitive classification to group the patterns. Counterpropagation works best when the patterns are tightly clustered in distinct groups.

Counterpropagation provides an excellent example of a network that combines different layers from other paradigms to construct a new type of network. Two different types of layers are used in counterpropagation: The hidden layer is a Kohonen layer, with competitive units that do unsupervised learning; the top layer is the Grossberg layer, which is fully interconnected to the hidden layer and is not competitive. The Grossberg layer is trained by a Widrow-Hoff or Grossberg rule.

OVERALL STRUCTURE AND DYNAMICS

Figure 10-1 shows the topology of a typical three-layered counterpropagation network. The first layer of units is solely for input, the second is the

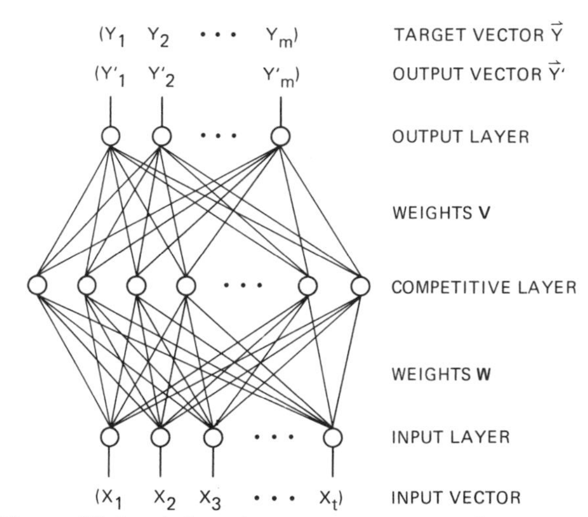

Figure 10-1. A three-layer counterpropagation network.

competitive — Kohonen — layer, and the third is the layer of output units (the Grossberg layer). Each layer is fully interconnected to the layer above it.

When trained, the counterpropagation network works as follows. First an input pattern is presented to the network (see Figure 10-2a). The hidden layer units sum their inputs and then compete to respond to that input pattern; a single unit wins the competition and becomes activated (see Figure 10-2b). The remaining hidden units become inactive.

When a hidden unit in a counterpropagation net wins the competition, this winning unit represents a classification category for the input pattern. Upon winning, the unit's activation level is set to 1, the activation levels of the other hidden units are set to 0 (see Figure 10-2b). The unit that wins the competition then activates a pattern in the top layer, which becomes the output of the network (see Figure 10-2c). Because the losing hidden units have activation values set at 0, they do not influence the activation of the output layer units. The winning unit, however, is fully connected to the output layer, and each interconnection has an associated weight. The values of those weights, thus, have a strong influence on the final values of the output units. The output pattern is read off the activation levels of the top layer units. The mathematical equations governing these activities are given in the following section.

BASIC EQUATIONS

Let us begin our discussion of the basic equations that the network uses when processing input patterns and adjusting its weights during learning by de-

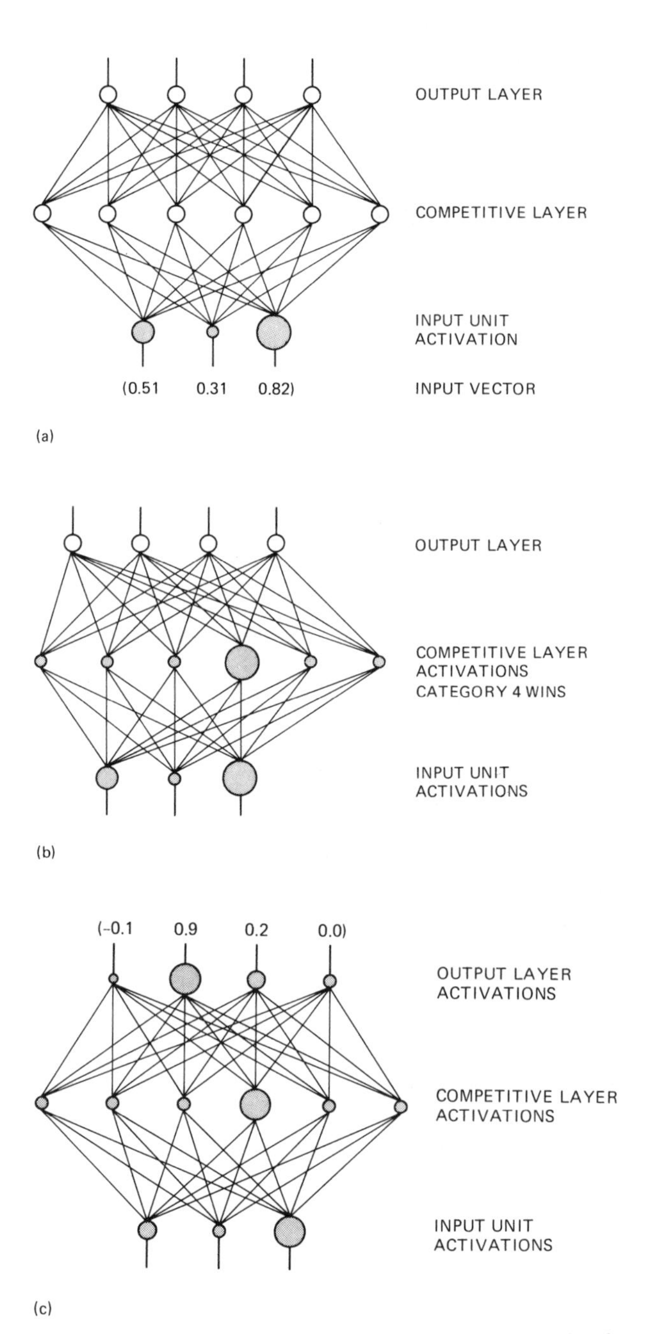

Figure 10-2. (a) An input vector is presented to the network. (b) A hidden layer unit wins the competition. (c) An output pattern is activated by the winning unit.

scribing the notation for an entire three-layered network, as shown in Figure 10-1. The input vector is

$$\mathbf{X} = (x_1, x_2, \ldots, x_t)$$

After competition, the activations of the competitive layer are

$$\mathbf{Z} = (z_1, z_2, \ldots, z_n)$$

which is a vector with 0/1 entries. The values for the network output are

$$\mathbf{Y}' = (y'_1, y'_2, \ldots, y'_m)$$

These are the activation levels of the output layer of units. The target output values are

$$\mathbf{Y} = (y_1, y_2, \ldots, y_m)$$

The network output $\mathbf{Y}'$ is intended to be an approximation of the target vector $\mathbf{Y}$. The training set consists of input-target pairs

$$(\mathbf{X}_a, \mathbf{Y}_a)$$
$$(\mathbf{X}_b, \mathbf{Y}_b)$$
$$(\mathbf{X}_c, \mathbf{Y}_c)$$

and so on.

The weights in the counterpropagation system are denoted as follows:

w_{ji} = weight from input unit i to competitive unit j

$\mathbf{W}_j = (w_{j1}, w_{j2}, \ldots, w_{jt})$ = weights incoming to competitive unit j

v_{kj} = weight from competitive unit j to output unit k

and

$\mathbf{V}_k = (v_{k1}, v_{k2}, \ldots, v_{kn})$ = vector of weights incoming to output unit k

The weight vector $\mathbf{W}_j$ is normalized. Thus,

$$\|\mathbf{W}_j\| = 1$$

The updating procedure preserves this normalization.

When input vector **X** is presented to the network, x_i becomes the activation level of unit i in the input layer. The hidden layer then performs the weighted sum

$$S_j = \sum_i x_i w_{ji}$$

where

$$S_j = \text{weighted sum for hidden unit } j$$

Competition then occurs in the hidden layer. Competition assigns a single unit in the hidden layer to be the winner. The unit with the highest sum S_j wins. Therefore

$$S_c = \max_i S_i \tag{10-1}$$

where the maximum is taken over all hidden units i. In the case of a tie, the unit at the left is chosen, by convention, as the winner. After competition, the hidden layer activations are:

$$z_c = 1.0 \tag{10-2}$$

$$z_i = 0.0 \text{ for all } i \text{ not equal to } c$$

After competition, the output layer does the weighted sum

$$y'_j = \sum_i z_i v_{ji} \tag{10-3}$$

The output for unit j is denoted y'_j. Because z_c is the only element z_i that is nonzero, and $z_c = 1$, (10-3) reduces to:

$$y'_j = v_{jc}$$

Thus each output unit takes on the value of the weight going from the winner to that output unit. (A variation on counterpropagation that involves more than one winner is described later.)

TRAINING

During training, both layers of weights are updated, the first after a competitive winner is chosen in the second layer of units. The competitive winner is chosen in response to the presentation of an input pattern. Only the weights of interconnections that go to the winning unit are adjusted, the others remain the same. After the competitive winner is selected, the output to the network is computed. The network's output values are then compared to the target pattern, and the second layer of weights is updated.

The incoming weights to competitive unit c are adapted as follows:

$$w_{ci}^{\text{new}} = w_{ci}^{\text{old}} + \alpha(x_i - w_{ci}) \tag{10-4}$$

where α is the learning constant $(0 < \alpha \leq 1.0)$. The other weights $w_{ji}(j \neq c)$ stay the same. Figure 10-3a illustrates the choice of weights to

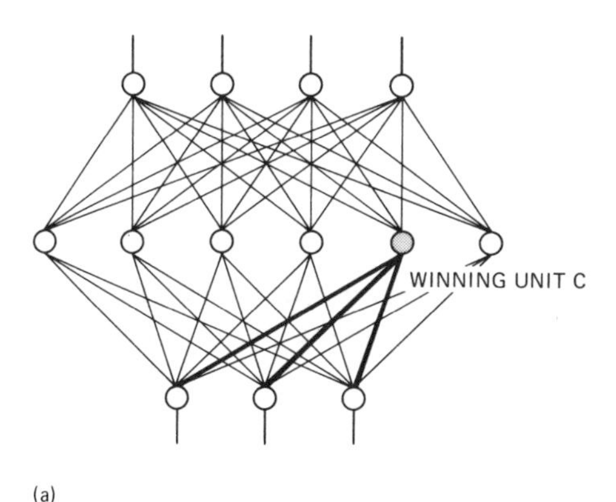

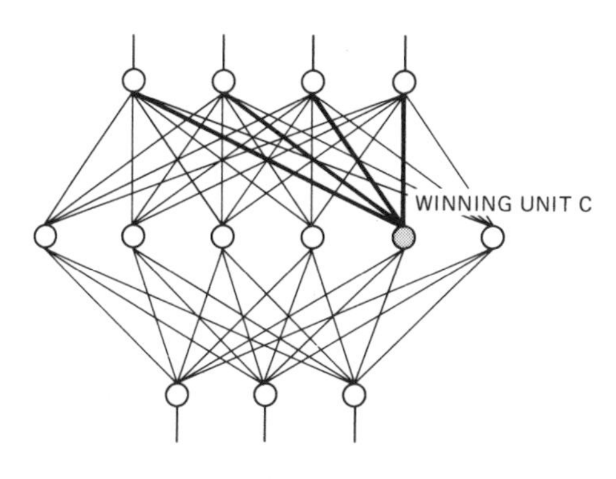

Figure 10-3. (a) Updating in the first layer of weights. Hidden unit c has won the competition; its incoming weights are then adjusted according to Equation (10-4). (b) Updating in the top layer of weights. Only the weights from the winner to the output layer are updated.

adapt. Thus, for $j \neq c$,

$$w_{ji}^{\text{new}} = w_{ji}^{\text{old}}$$

After adaptation rule (10-4) is employed, the weight vector $\mathbf{W}_j$ is renormalized by being divided by its current length:

$$\mathbf{W}_c^{(\text{normalized})} = \mathbf{W}_c / \|\mathbf{W}_c\|$$

Adaptation of the output layer weights is done by the Widrow-Hoff rule (Hecht-Nielsen 1988):

$$v_{ji}^{\text{new}} = v_{ji}^{\text{old}} + \beta z_i (y_j - y_j') \qquad (10\text{-}5)$$

Only one competitive unit is active at a time. For the winning unit $z_c = 1$; all other units are set to 0. As a result, Equation (10-5) reduces to

$$v_{ji}^{\text{new}} = \begin{cases} v_{ji}^{\text{old}} \text{ if } i \neq c \\ v_{ji}^{\text{old}} + \beta (y_j - y_j') \text{ if } i = c \end{cases}$$

Thus the only weight adjusted for output unit i is the weight connected to the winning unit c. Figure 10-3b shows the choice of adapted weights going to the output layer.

VECTORS AND NORMALIZATION

The basic counterpropagation network requires the input vector to be normalized. Normalization means that the length of the vector is 1.0:

$$\|\mathbf{X}\| = 1.0$$

where

$$\|\mathbf{X}\| = \sqrt{\left(\sum_i x_i^2\right)} = 1$$

With normalization, the input vectors lie on a unit sphere (or hypersphere, if vectors have more than three entries). Figure 10-4a illustrates a set of input vectors in three dimensions that lie on the unit sphere.

The weight vectors

$$\mathbf{W}_j = (w_{j1}, w_{j2}, \ldots, w_{jn})$$

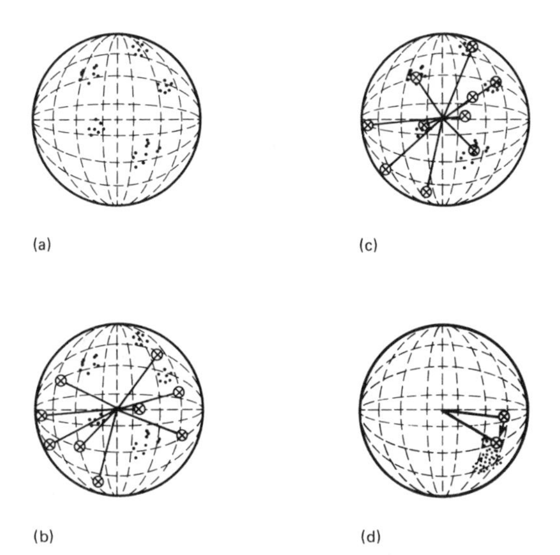

Figure 10-4. (a) Input vectors are arranged in clusters on a unit sphere. (b) Weight vectors are randomly placed on the sphere before training begins. (c) After training, many weight vectors are located inside the clusters of input patterns. (d) The weight vector $\mathbf{W}_c$ for the winning unit is moved a step closer to the current input pattern, X.

to the competitive layer are also normalized. Thus, the weight vectors $\mathbf{W}_j$ lie on a unit sphere (or hypersphere), also. Figure 10-4b shows an initial placement of weight vectors; Figure 10-4c shows a final placement after training is complete.

The incoming sum to a hidden unit reflects the angle between the input vector and the weight vector. The reason for this is the following mathematical relationship. The incoming sum to competitive unit j is

$$S_j = \sum_i w_{ji} x_i$$

which is equal to the dot product

$$\mathbf{W}_j \cdot \mathbf{X} \tag{10-6}$$

But from standard vector relationships the dot product is

$$S_j = \|\mathbf{W}_j\| \, \|\mathbf{X}\| \cos \theta_j \tag{10-7}$$

where θ_j is the angle between $\mathbf{W}_j$ and $\mathbf{E}$. An explanation of (10-6) and (10-7) can be found in most textbooks on vector calculus (Shenk, 1979, p. 504). Since $||\mathbf{W}|| = ||\mathbf{X}|| = 1$, we have

$$S_j = \cos\theta_j$$

Then c is such that

$$\cos\theta_c = \max_i(\cos\theta_i)$$

or

$$\theta_c = \min_i \theta_i$$

But θ_c is the angle between the input vector and the weight vector for the winning unit. Thus the winning unit has the smallest angle between its weight vector $\mathbf{W}_j$ and the input vector $\mathbf{X}$.

The weight adjustment specified in (10-4) has the effect of moving $\mathbf{W}_c$ toward the input vector $\mathbf{X}$. Figure 10-4d illustrates this move geometrically. The next input vector for which the same unit c wins will move the weight vector $\mathbf{W}_c$ again, but the vector will then move in a direction toward the new input vector. In the long-term, the weight vector usually approximates an average of all the input vectors for which it won. As successful learning occurs, the processing units in the competitive layer adjust their weight vectors to divide up the input space in approximate correspondence to the frequencies with which the inputs occur. Figure 10-4c represents the trained network. The clusters in the figure have approximately the same number of input points in each, and one weight vector is central to each cluster.

The $\mathbf{W}_j$'s are considered to be *exemplar* (model) vectors. A weight vector $\mathbf{W}_j$ in the trained network can be an exemplar vector to a cluster or group of input vectors. A given input pattern activates the exemplar vector that is closest to the input pattern. It is the winning unit's exemplar vector $\mathbf{W}_c$, then, that causes the win. A single output pattern is activated for each unit that wins in the hidden layer.

HOW TO DO NORMALIZATION

Normalization is required when pattern vectors to be used as inputs do not have the length of 1. This situation often arises when real data are used for the patterns that are presented to the network. Normalization involves changing a pattern vector so that the length of the input vector is 1 when it is presented to the network. A number of possibilities of how to do normalization (NeuralWorks 1988) exists. The choice depends on which changes can be made to the data without interfering with their general properties. We demonstrate two methods. One is retain the whole vector but add an entry and scale the vector, the other, just scale the vector. The latter is appropriate only when the relative values of the entries are important but the total length of the vector is not.

Method 1

Start with a data vector $\mathbf{E}'$. Assume that this vector represents the pattern to be input; however, it is not normalized. Thus

$$(e_1, e_2, \ldots, e_n) = \mathbf{E}'$$

Choose a value N that is slightly greater than the length of the longest input vector in the application under study. Add a new entry to the input vector:

$$(d, e_1, e_2, \ldots, e_n) = \mathbf{E}''$$

Now set

$$d = \sqrt{(N^2 - \|\mathbf{E}'\|^2)}$$

Now divide the vector $\mathbf{E}''$ by N to get a new vector $\mathbf{E}$ of length 1:

$$\mathbf{E} = \frac{\mathbf{E}''}{N}$$

The vector $\mathbf{E}$ is normalized, and may be used as input to the network.

Figure 10-5a illustrates this method for the vector

$$\mathbf{E}' = (e_1, e_2) = (0.7, 0.3)$$

shown plotted on the x-y plane in the figure. The third entry, d, is plotted in the third dimension. The length of d has been constructed to be exactly long enough to place the end of the vector on a sphere of radius N. The vector is then divided by N, and the vector becomes placed on a sphere of radius 1.

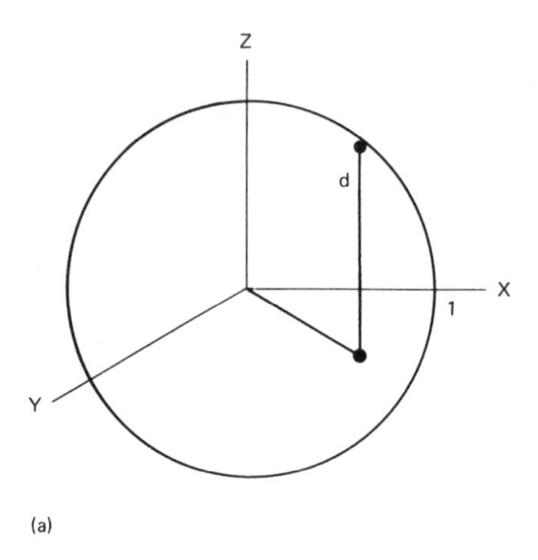

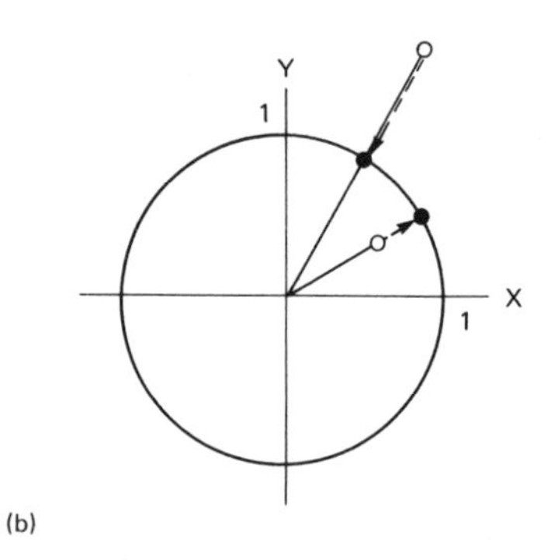

Figure 10-5. (a) Normalization is accomplished by adding an entry to the data vector. (b) Normalization is accomplished by scaling the data vector.

Method 2

With this method (illustrated in Figure 10-5b), the original pattern is shortened or lengthened to be on the unit sphere, and is recommended mainly for situations in which the direction of the data vector is important, but its total length is not. The direction of the vector is determined by the relative values of the different entries in the vector. Thus the method is best used when the relative values of the vector entries are more important than the total length of the vector. The following equation computes the normalized vector:

$$\mathbf{E} = \frac{\mathbf{E}'}{\|\mathbf{E}'\|}$$

where $\mathbf{E}'$ is the data vector and $\mathbf{E}$ is the normalized vector. The resulting vector $\mathbf{E}$ has length 1, and is used as input to the network.

A PATTERN CLASSIFICATION EXAMPLE

Figure 10-6 presents a pattern classification problem in which the pattern classes are clearly separable. In this example, patterns consisted of two-dimensional vectors that were graphed as points on the plane. Pattern classes consisted of four different clusters of points, each of which was spread across a square of diameter 0.1. Figure 10-6a shows the counterpropagation network trained to classify these patterns. There were 30 units in the hidden layer, which classified the data in its unsupervised learning mode. During training, some of the hidden units won and changed their weights to be centered in the clusters; these became exemplar vectors. Other hidden units did not win, and remained in the same place in which they were randomly positioned at the start.

The plotted points from the original data are given in Figure 10-6b. Each data vector was normalized to a three-dimensional vector using Method 1. The network's exemplars thus had three entries, the first two of which are plotted on the graph in Figure 10-6a. The values of these entries have been rescaled to match the original data vectors. Four exemplar vectors were developed during training; each became positioned near the center of the cluster to which it responded best.

The top layer of processing units divided the pattern clusters into four classes. Each output unit signified a different category. A portion of the test set is shown in Figure 10-6c. After 1,000 iterations with both learning constants set at 0.1, the network classified with 100% accuracy.

AN ADVISORY NETWORK EXAMPLE

In this fanciful network adapted from NeuralWorks (1988) the user has a choice of what to do on Sunday afternoon, and builds a neural network to help advise him. Although this task can be implemented easily with a rule-based

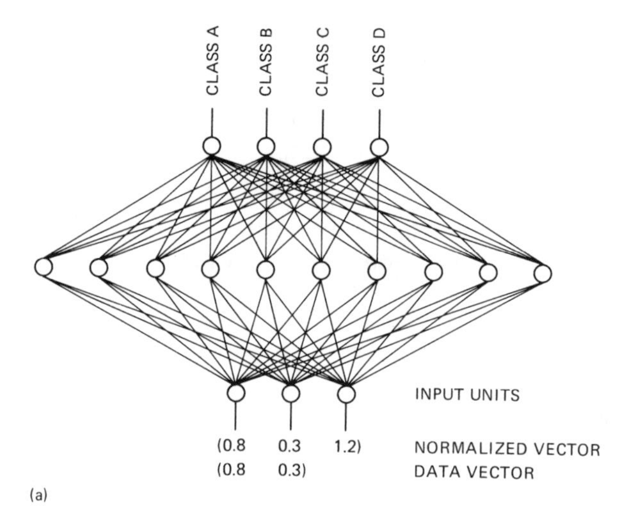

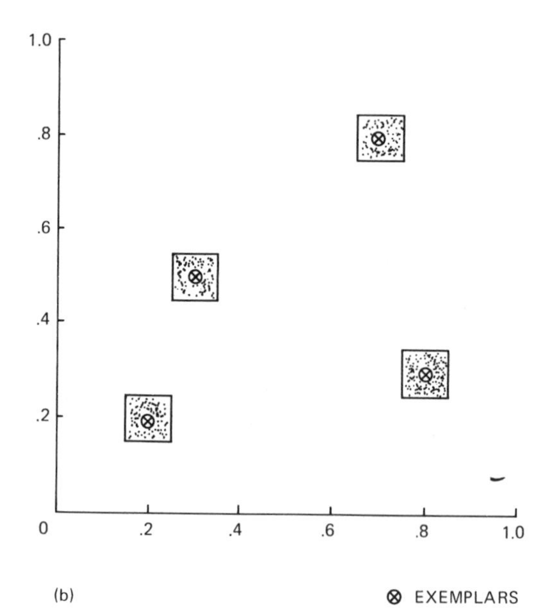

Figure 10-6. (a) Counterpropagation network trained to classify pattern clusters. (b) Plotted points from original data and exemplar vectors from the trained network. Data vectors are uniformly spread across the four square areas in the figure. Exemplar vectors are shown in each area.

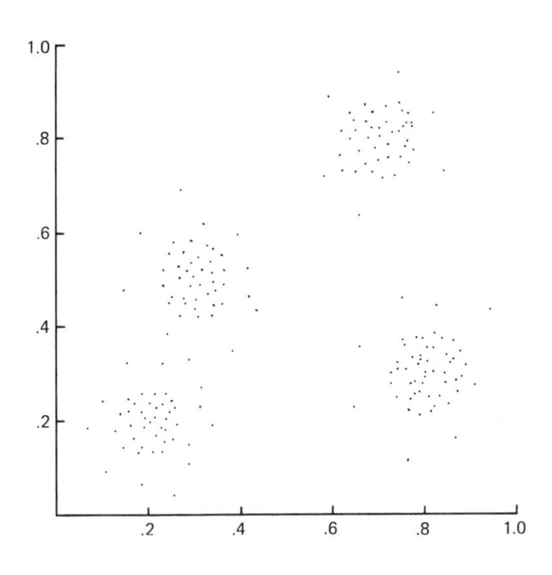

Figure 10-6. (cont.) (c) Part of the test set used to verify that the network in 10-6a was classifying the patterns in 6b correctly.

expert system, this example shows how a neural network might be used. The possible activities are: work, watch cartoons, go shopping, walk in the park, and have dinner with a friend. Factors to take into account are the amount of work one has to do at the office and one's current level of sentimental feeling.

Figure 10-7a shows the training set and Figure 10-7b shows the network used for this example. In Figure 7b the network is responding to the input (0.5, 1.0). A single hidden unit is activated (unit 11), and in turn this unit activates the output unit that recommends the activity of "dinner."

After training, the network classifies the training set perfectly. The trained network also has an "interpolation" or "generalization" mode. In this mode, the network is presented with inputs that are gradations between the values of 0.0, 0.5, and 1.0 that appeared in the training set. For example, Figure 10-7c shows the response of the network to a set of input patterns that hold "level of feelings" fixed at 1.0 but vary "amount of work" between 0.5 and 1.0. A transition point at 0.8–0.9 is reached, where the network changes from recommending "dinner" to recommending "walk in park." The network interpolates as an advisor. In this case it has found a reasonable transition point between the two alternative activities. This type of interpolation (or generalization) is not unique to counterpropagation, and can be found in other networks such as back-propagation as well. Generalization of this sort

AMOUNT OF WORK	LEVEL OF FEELINGS	ADVISED ACTIVITY
NONE (0.0)	LOW (0.0)	WATCH CARTOONS
SOME (0.5)	LOW (0.0)	WATCH CARTOONS
NONE (0.0)	MEDIUM (0.5)	GO SHOPPING
LOTS (1.0)	HIGH (1.0)	WALK IN PARK
SOME (0.5)	HIGH (1.0)	DINNER
LOTS (1.0)	MEDIUM (0.5)	WORK

(a)

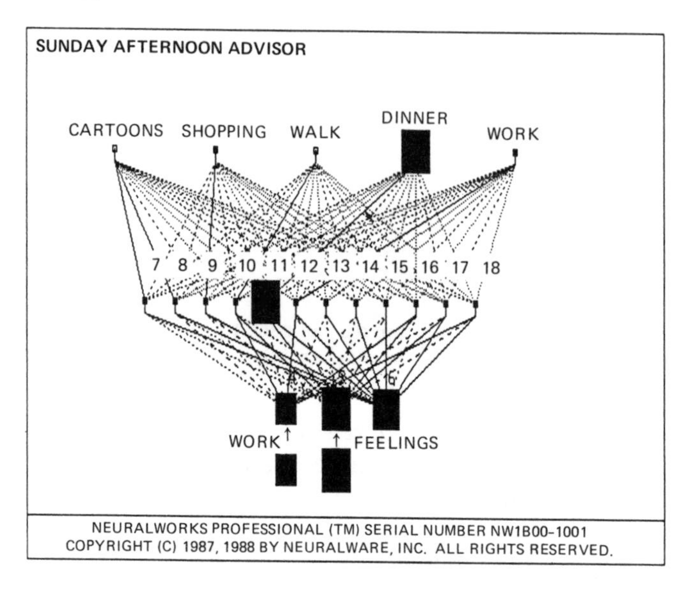

(b)

AMOUNT OF WORK	LEVEL OF FEELINGS	ADVISED ACTIVITY
0.5	1.0	DINNER
0.6	1.0	DINNER
0.7	1.0	DINNER
0.8	1.0	DINNER
0.9	1.0	WALK IN PARK
1.0	1.0	WALK IN PARK

(c)

Figure 10-7. (a) Training set for Sunday afternoon advisor. (b) The network for Sunday afternoon. Each square represents a processing unit, and activation levels are shown by the sizes of the squares (From *Neural Works Professional II Manual.* Neuralware 1988.) (c) A range of inputs and responses for the advising network. "Amount of work" is varied from 0.5 to 1.0, while "feelings" are fixed at 1.0.

must be tested carefully to insure that the network has satisfactory performance in each real-world application.

VARIATIONS AND IMPROVEMENTS

Training Schedule

A number of variations can be made to the standard counterpropagation training rules. First, a network can be trained in two steps: In the first step, only the bottom layer of weights is adjusted. Many training iterations are done until the competitive layer has developed a classification scheme for the input patterns. In the second training step, the second layer of weights is adjusted also. The rationale behind this method is that the second layer of weights does not develop meaningful weights until the first layer of weights has undergone some training.

Another variation is to change the values of the training constants. They may begin large and become smaller as the training iterations progress. This allows larger changes to occur first, followed by refinement in the values of the exemplar vectors and the output patterns.

Recall With Two Winners

The standard counterpropagation network allows there to be exactly one winner in the competitive layer. An alternative structure is limited to one winner in the competitive layer only during training; thus, the training of the network is the same as we have already described. Recall mode, however, is different in that more than one winner is allowed in the competitive layer. Usually two such winners are allowed, and are chosen as the two hidden units with the highest incoming sum. After competition, each of these units sets its activation level to 1, with the other units set at 0. The two winners both influence the weights of the output layer according to Equation (10-3). The output activations become a sum of the activations caused by each of the two winning units. Figure 10-8 shows an example of this type of behavior with a simplified training set. We have chosen a simple example here, where the two inputs do not interfere with one another's category units. In more complex examples, performance must be verified through experimentation with real data.

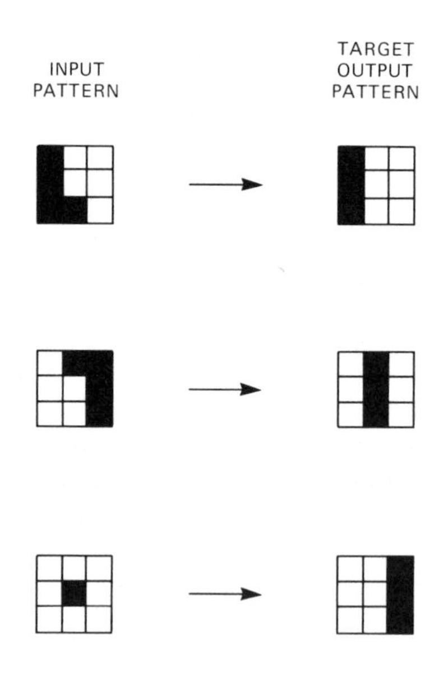

Figure 10-8. (a) Simple training set for a counterpropagating network.

Top-Down Influence

The counterpropagation network does not communicate information from the output or target values to the first layer of weights. It is thus possible for the network to make a mistake by categorizing a set of similar input patterns into the same group when they should belong to two separate groups. Figure 10-9 shows such an example, with weights assigned for illustration. The training set is given; the competitive layer classifies two input patterns with the same hidden unit when they are supposed to map to two very different output patterns. To address this issue, top-down information that forces the network to assign different competitive units to input patterns that map to different target patterns can be supplied. A number of other network designs attempt to overcome this type of problem (Carpenter and Grossberg 1988; NeuralWorks 1988; Reilly et al, 1982).

Counterpropagation Without Normalization

Counterpropagation networks can be built without the restriction that the input vectors be normalized (Hecht-Nielsen 1988). Counterpropagation without normalization uses a competitive rule that is the same as in the

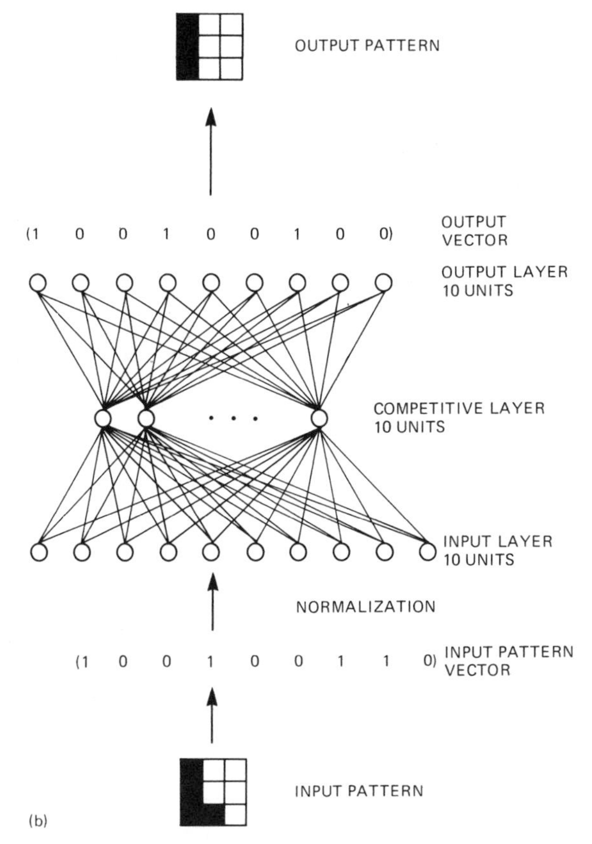

Figure 10-8. (cont.) (b) The counterpropagating network trained. Training mode here allowed exactly one winning unit in the hidden layer for each input pattern presentation. The network used did not require normalization.

Kohonen feature map described in Chapter 9. The distance between two vectors (input vector $\mathbf{X}$ and exemplar $\mathbf{W}_j$) is calculated for each hidden unit j. The unit with the smallest such distance is the winner.

The procedure is to choose winner c such that

$$\|\mathbf{X} - \mathbf{W}_c\| = \min_j \|\mathbf{X} - \mathbf{W}_j\|$$

The output layer activations are then calculated the same way as described in Eqs. 10-1 through 10-3, and the adaptation rule is the same as Eq. 10-5. As an example, the calculations in Figure 10-6 can be replicated with a network that does not require normalization.

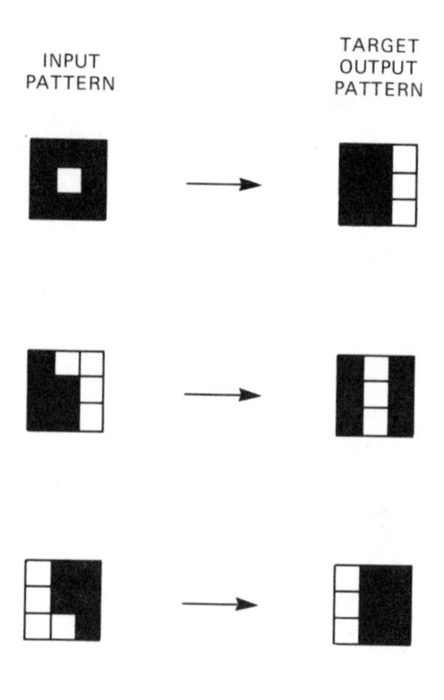

Figure 10-8. (cont.) (c) Some results of recall mode using two winning units instead of one in the hidden layer. If the sum of two training patterns is input, then the sum of the two corresponding target output patterns are produced by the network.

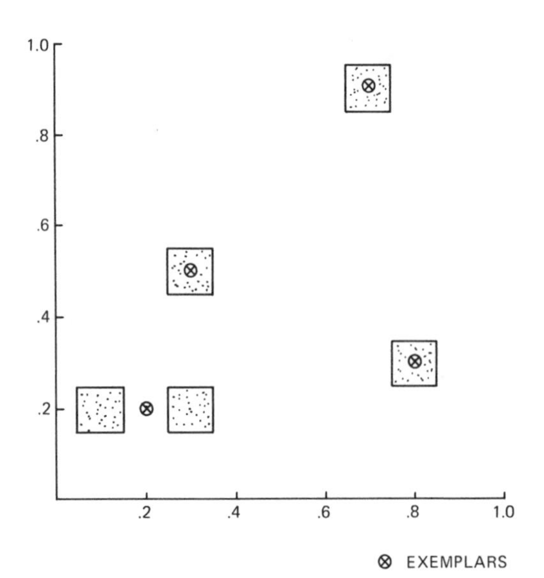

Figure 10-9. Weight vector exemplars developed by training a counterpropagation network. The exemplar at the bottom left responds to two clusters.

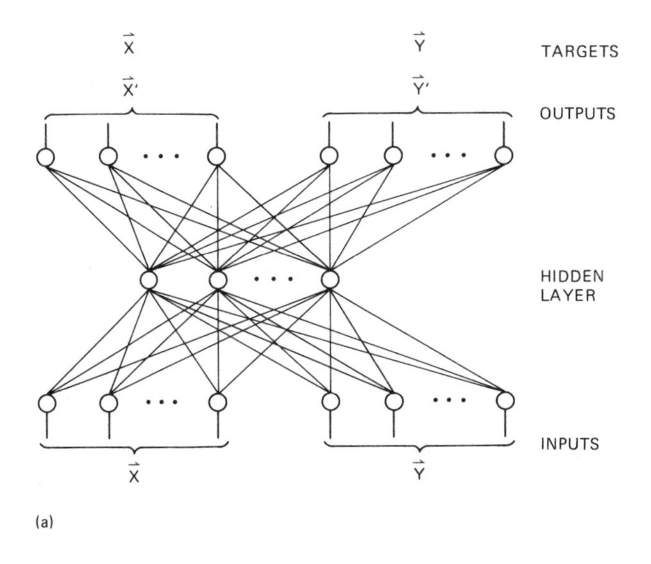

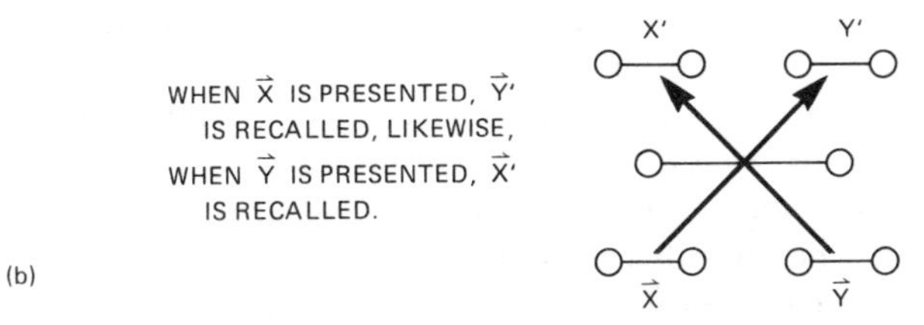

Figure 10-10. (a) A five-layered counterpropagation system. (b) The flow of activations is shown for presenting *X* to recall *Y*, and for presenting *Y* to recall *X*.

The "Five-Layered" Counterpropagation Network

The term counterpropagation comes from the network arrangement wherein such a net is used as an associative memory: Activations are depicted as though they cross one another, hence the name counterpropagation. An associative memory can be set up as in Figure 10-10, where the inputs are the dual vectors $(\mathbf{X},\mathbf{Y})$ and target outputs are the same pair of vectors $(\mathbf{X},\mathbf{Y})$.

The network is then trained on a training set:

In	Out
$(\mathbf{X}_1,\mathbf{Y}_1)$	$(\mathbf{X}_1,\mathbf{Y}_1)$
$(\mathbf{X}_2,\mathbf{Y}_2)$	$(\mathbf{X}_2,\mathbf{Y}_2)$
$(\mathbf{X}_3,\mathbf{Y}_3)$	$(\mathbf{X}_3,\mathbf{Y}_3)$

and so on. If training is successful, then associative recall may be done by presenting the network with the vector **X** and having it output the combination **X**′,**Y**′, which is an approximation to (**X**,**Y**). Thus, **X** produces an approximation to **Y**. Likewise, **Y** may be presented to produce an approximation to **X**.

APPLICATION: DOLPHIN ECHOLOCATION

A counterpropagation network has been used to duplicate some of the echolocation abilities of the dolphin. In a study by Roitblat and colleagues (1989), a neural network was given as input the echo returns typically used by dolphins as a means of locating and identifying objects in the water. The performance of the trained network was excellent. The task was to classify four different underwater targets; the network performed at 100% accuracy in a simulated noise-free environment and at 97% in a realistic noisy environment. Dolphins were able to attain a 95% accuracy in the same noisy natural environment.

Dolphins have evolved unique sonar capabilities for detecting and recognizing objects in natural environments. They frequently swim in inlets and bays that are so murky that vision is severely restricted; sonar provides them with a system for detecting objects in the water. With sonar, dolphins can discriminate between objects differing in size, structure, shape, and material composition. The dolphins perform this discrimination by issuing a high-frequency click and then evaluating the return echo. The return time depends on the distance of the animal from the target it is scanning. Typically there is a short gap between the echo return from one click and the emission of the next.

Four different targets were used: a cone, a sphere, a large tube, and a small tube. In the simulated environment, no activity was going on in the water, and an artificial sound generator was used to make a click similar to a dolphin's click. In the natural environment, the dolphin was swimming in a water tank. The dolphin generated its own clicks and a sensor attached to the dolphin recorded the return signals.

In each case, the return echoes were digitized and subjected to a Fourier transform. Figure 10-11 shows an example of the amplitude and Fourier transform for each item. The center portions of the Fourier transforms were divided into 20 frequency bins, and average height was calculated for each bin. These values furnished a 20-dimensional data vector for the neural network.

Normalization was done on the data vector according to Method 1. The normalized vector then had 21 entries, and was used as the input vector to the

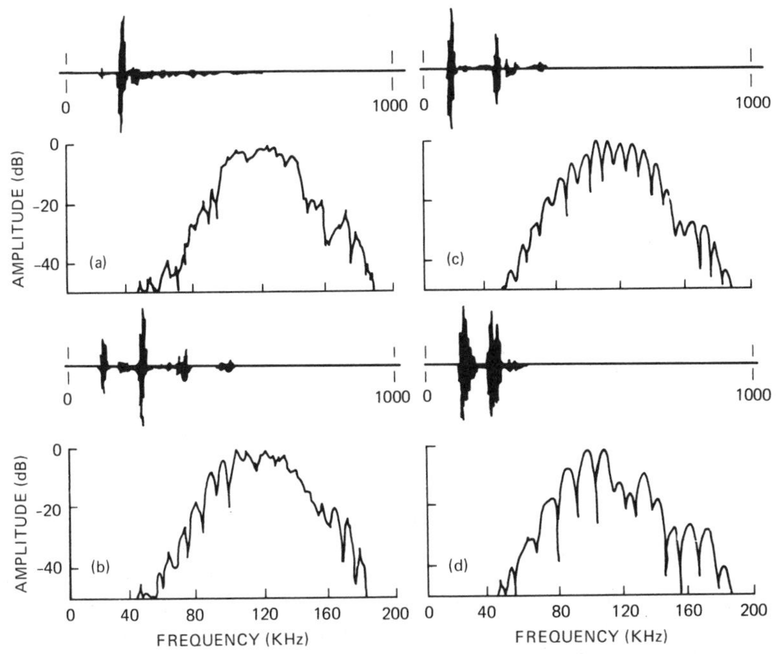

Figure 10-11. Amplitude display and Fourier transform for each item in the dolphin echolocation experiment. (From H. L. Roitblatt, et al. Dolphin Echolocation: Identification of returning echoes using a counterpropagation network. *Proc. IJCNN.* © 1989 IEEE.)

counterpropagating neural network. The competitive layer consisted of eight units, and the output layer had four, with one unit to identify each of the target categories. The network was trained with a total of 5,000 iterations.

Figure 10-12 summarizes the average amount of dissimilarity among the different signals. A root-mean-square value was calculated to reflect differences among patterns in the same category, and differences among patterns in different categories. The averages for within-category distances are given on the diagonal in the table, and the between-category distances are off the diagonal. The values for between-category distances were consistently higher; this reflects the fact that the target categories were discernable.

The top of Figure 10-12 is based on the natural noisy environment, and the bottom of Figure 10-12 is based on the simulated noise-free environment. The between-category distances were still higher in the noisy environment than the within-category distances, however the difference was not as great as in the noise-free environment. The dissimilarity distances indicate that discrimination is more difficult in the natural environment.

This study demonstrates the success of a counterpropagation neural net-

AVERAGE DISSIMILARITIES

RELATIONS AMONG CATEGORIES OF NATURAL ECHOES

	CONE	*SPHERE*	*L TUBE*
CONE	0.84	1.86	2.39
SPHERE		1.24	2.54
LTUBE			0.84

RELATIONS AMONG CATEGORIES OF ARTIFICAL ECHOES

	CONE	*SPHERE*	*L TUBE*	*S TUBE*
CONE	0.08	1.57	1.66	3.79
SPHERE		0.08	1.16	3.47
L TUBE			0.26	2.71
S TUBE				0.09

Figure 10-12. Differences within and between categories for the four echolocation items. The top table is for the simulated noise-free environment and the bottom table is for the natural, noisy environment. (From H. L. Roitblatt, et al. Dolphin echolocation: Identification of returning echoes using a counterpropagation network. *Proc. IJCNN.* © 1989 IEEE.)

work at recognizing a set of sonar targets — cone, sphere, and small and large tubes. The discrimination task was relatively simple, as the targets were chosen to be readily discriminable. The performance level, however, was exceedingly good, reaching 97% – 100% as compared to the 95% accuracy of the dolphins.

OVERVIEW

Counterpropagation is considered a faster alternative to back-propagation. The improvement in training time with counterpropagation is usually substantial. However, a question remains about the performance. Although counterpropagation can learn many pattern-mapping problems well, studies have shown that it often generalizes less well on new patterns (Glover 1988).

Counterpropagation requires that input pattern classes be organized into clusters that are separated (nonoverlapping). Inputs should uniformly cover class categories — irregular concentrations of input vectors may skew the concentrations of elements in the competitive layer. Faulty divisions of pattern classes can result from this skewing. Counterpropagation requires enough processing units in the competitive layer to be able to adapt to the boundaries between pattern classes. Too few competitive units lead to coarse contours in the boundaries.

A critical flaw in counterpropagation networks is that occasionally the competitive layer can become unstable. Instability results when there are not enough hidden units to represent the pattern groups. In this case, a single hidden unit responds to patterns in more than one class, and can actually migrate from representing one class to representing another class indefinitely. If more hidden units are added (a sufficient number), then the network usually reaches a stable state in which hidden units are dedicated to pattern groups, and do not change their affiliations.

Candidate applications of counterpropagation include the entire gamut of pattern-mapping problems that have been identified from such diverse fields as medicine, aerospace, and finance (see Chapter 11). Particular applications areas that have been proposed for counterpropagation include pattern classification, statistical clustering, and data compression. Notable studies have been done on industrial inspection (Glover 1988) and sonar return classification (dolphin signals) (Roitblat et al. 1989).

The calculations of a counterpropagation network do not require a neural network implementation; these basic calculations can be expressed with mathematical notation and relatively simple algorithms. However, the counterpropagation network does organize a parallel implementation for its pattern lookup and associative memory capabilities. Thus there are possibilities for accelerated hardware design that take advantage of the parallel nature of the network. Counterpropagation networks are adaptable in the sense that one can continually retrain them to respond to changing circumstances. In this case, the exemplar vectors will readjust to changes in the characteristics that distinguish the data classes. A variety of variations to counterpropagation are available, each adding additional power to the capabilities of the paradigm.

References

Carpenter, G. and Grossberg, S. 1988. *Computer,* March 1988:77–88.

Glover, D. 1988. A hybrid optical/electronic neurocomputer machine vision inspection system. Procedures of Vision 88 Conference, Society Manufacturing Engineers, Dearborn, MI.

Hecht-Nielsen, R. 1988. Applications of counterpropagation networks. *Neural Networks* 1(2):131–140.

Hecht-Nielsen, R. 1987. Counterpropagation networks, *Proc. of IEEE First Int'l Conference on Neural Networks. 1987* II:19–32.

NeuralWorks Professional II Manual. 1988. Sewickley, Pa.: Neural Ware Inc.

Reilly, D. L., Cooper, L. N., and Elbaum, C. 1982. A neural model for category learning. *Biological Cybernetics* 45:35–41.

Roitblat, H. L., Moore, P. W. B., Nachtigall, P. E., Penner, R. H., and Au, W. W. L. 1989. Dolphin echolocation: Identification of returning echoes using a counterpropagation network. *Proc. of the International Joint Conference on Neural Networks, 1989* I:295–299.

Shenk, A. 1979. *Calculus and Analytical Geometry,* Second Edition. Santa Monica, California: Goodyear Publishing Company.

11

Applications and Future Directions

In this book we have described a wide variety of neural network paradigms, shown their basic activation dynamics, and covered the abilities and applications for each. In this final chapter, we take an overview of the many applications of neural networks by looking at the network as a pattern-mapping system that could be placed in a wide variety of domains — medicine, manufacturing, image systems, speech systems, autonomous control, and diagnostics, among others. We also discuss some key concerns and design issues in the development of new applications for neural network technology.

The word "applications" should be treated with care. In evaluating an application, it is important to be aware of how well developed an application is at the time of evaluation. The most preliminary category of applications — *candidate applications* — are those problems that could, in principle, be solved by the type of technology that neural networks offer. In a candidate application, the problem has been identified and the problem requires a mapping or optimization that is similar to other problems addressed by neural networks. A candidate application problem, however, has not yet been solved by a neural network. Further work is required to develop a neural network to solve the problem, and thereby to prove that a neural network is adequate to the task.

Applications under development are those problems for which studies have been performed or are underway. In these studies, a neural network is usually trained to learn a simplified version of the problem first, and then expanded to address the entire problem. A complete solution, however, has not necessarily been found. Most neural network applications today (as of the writing of this book) can be considered under development.

Proven applications are those for which neural networks are actually used

The three-tiered organization of applications:

1. Candidate applications
2. Applications under development
3. Proven applications

Figure 11-1.

to solve problems. Such proven applications are often commercially distributed. The first proven application of neural networks was in handwritten character recognition; neural networks were trained to read handwritten characters in spite of the inherent variations. Another proven application is in bomb detectors (Shea and Lin 1989) where a neural network is used to classify sensor patterns to ascertain whether an explosive is present.

Although there is some overlap in these three levels of applications (Figure 11-1), the classifications are useful. Exaggerated claims about neural networks commonly point to many applications for neural network technology but fail to mention the stage of development for each application. Although the technology has far-reaching impact, it is important to evaluate how much work is yet needed to bring each application to fruition.

When discussing applications, a further division should be made between the task of identifying where a successful neural network system could be useful and the task of building the neural network itself. In most applications areas, it is not a trivial issue to identify which unsolved problems could be addressed by a neural network. Once the appropriate problems have been identified, it is then a separate task to build a network that could in fact produce the needed outputs when presented with the input data.

For the sake of illustration, suppose there were a magic black box that could solve any pattern-mapping problem (Figure 11-2), and suppose that part of the magic of the box is its ability to infer the exact pattern map that is needed.

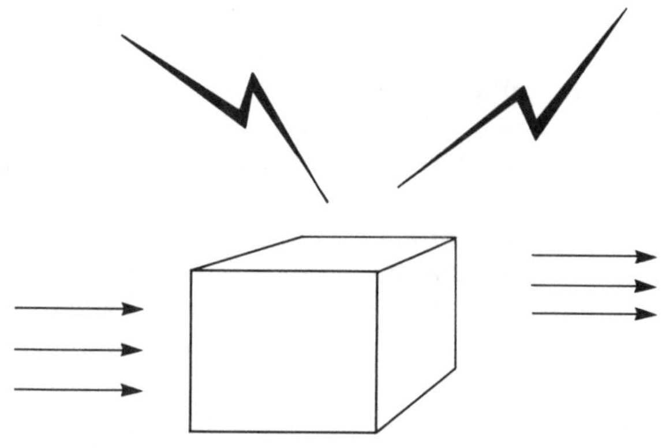

Figure 11-2. The magic black box for pattern-mapping.

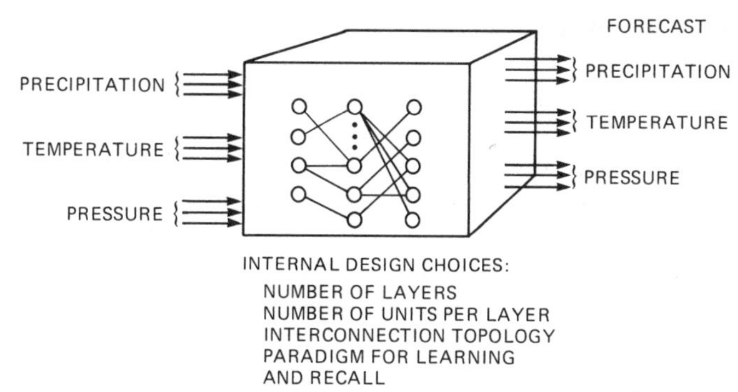

Figure 11-3. Basic scheme for a weather forecasting network. Design choices for the neural network are listed.

The task of identifying a candidate application for a neural network is the task of finding where this black box could be used. The task of actually building a network that performs successfully comes second.

Early work by Widrow entailed many ideas for where an adaptive pattern-mapping system could be useful (Widrow and Stearns 1985). Although his neural networks were based on simple adaline-like systems, he visualized a wide scope of places for the use of a pattern-mapping/pattern-classification tool. He attempted applications in speech (syllable) recognition, weather forecasting, two-dimensional pattern classification, adaptive control, and adaptive noise filtering. He was limited by the two-layered architecture of the day; however, his identification of candidate applications was a lasting contribution. Today we can build more sophisticated neural networks to perform the tasks that were identified earlier.

Figure 11-3 illustrates a neural network "black box" for pattern mapping. Interfacing to the box are a variety of input and output channels. For each channel, information is represented in a specific way, with the particular representation chosen by the designer. Internal to the box are the design choices for the neural network. These design choices include the size of the network (how many layers and how many units in each layer), the interconnection topology, and the paradigm used for learning and recall. The neural network is trained on available data, and after training is tested to evaluate performance. Performance can be compared for different network designs to identify the best network for the job.

Using weather forecasting as an example, let the first few input channels code for the amount of rain today, let the next set of channels code for the temperature, and let the last set of channels code for the pressure (as in Figure 11-3). The outputs are the same measurements, but represent forecasts for tomorrow's weather. A past history of weather data is used to train the network. In the training set, each day's weather is an input pattern, and the following day's weather is the corresponding target output pattern. After training, the network is put into recall mode and given today's weather parameters. It outputs a forecast for tomorrow.

Many network designs could be used for the weather-forecasting problem. Figure 11-3 shows a three-layered network. A number of different paradigms, network sizes, and topologies could be used with the same training data on weather. A neural network study in this case would assess which network performed better over a variety of paradigms and network sizes.

SPECIFIC APPLICATIONS

Candidate applications for neural networks arise in any situation that can benefit from an ideal pattern-mapping system. Problems that require pattern completions and pattern classifications can be addressed by neural networks. Sometimes optimization tasks can be addressed. Myriad specific tasks fall into these categories: An example pattern-mapping network would input written text and output a representation of a spoken word, an example pattern-completion network would fill in details on a partially obscured picture by recalling a stored pattern, an example pattern-classification network would sort different images into categories, an example optimization network would arrive at a preferred layout pattern for an electronic circuit. These examples illustrate the nature and breadth of potential neural network applications.

Financial Analysis

The financial world presents an opportunity to use neural networks to improve earnings and lower losses. Candidate applications of neural nets include financial evaluation, forecasting, and analysis. A neural network can behave as a nonlinear financial predictor, taking as input financial data and producing as output a financial assessment or forecast. A neural network can be trained to evaluate loan or insurance information or to evaluate other complex financial situations.

A simple financial model would take as inputs the past history of financial data such as stock prices, financial indicators, and financial statistics (Figure 11-4). The network outputs would forecast different financial assessments or outcomes. Perhaps such a system would take as input the past history of stock prices and its output would be a forecast of future stock prices. For a more

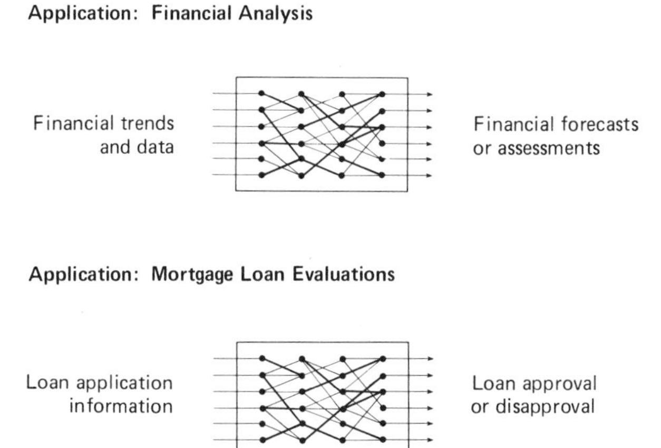

Figure 11-4. Levels of applications development.

successful result, however, a more sophisticated set of inputs with a variety of sources for pertinent financial data would be needed (Collins, Ghosh, and Scofield 1988). Neural networks have been researched as tools in portfolio management and financial forecasting.

Neural networks have also been trained to perform mortgage loan evaluations (see Figure 11-4) by companies that finance mortgages and companies that insure mortgages. For both, it is important to predict the risk of mortgage loan default based on information given in the mortgage application. Loan applications include information on the borrower's personal status, income, obligations, the mortgage amount, and the property. A hundred or more such items of information are evaluated for each loan application. Neural networks have been applied to input this information and output a risk evaluation. Inputs and outputs for one such study are summarized in Figure 11-5 (Collins, Ghosh, and Scofield 1988). The neural networks were trained on a pool of mortgage applications. Input patterns in the training set represented the information given in each loan application. The target outputs in the training set were the outcome of each loan — whether or not the applicant defaulted. The trained network then could produce a decision as to whether or not to grant a loan. Good performance was reported from this approach (Collins, Ghosh, and Scofield 1988) with a system that contained 6,000 processing units. Multiple neural networks were used, each of which performed part of the task of application evaluation. The neural networks appeared to outperform the underwriters' judgments, which could result in substantial financial savings to the loan industry.

INPUTS

BORROWER'S "CULTURAL" STATUS:

CREDIT RATING
NUMBER OF DEPENDENTS
NUMBER OF YEARS EMPLOYED
SELF-EMPLOYED
INTENDED TO OCCUPY PROPERTY

BORROWER FINANCIAL STATUS

CURRENT INCOME
PORTION OF INCOME DUE TO SOURCES OTHER THAN SALARY
AMOUNT OF OBLIGATIONS OTHER THAN THE PRINCIPAL PROPERTY

MORTGAGE INSTRUMENT

LOAN-TO-VALUE RATIO
TYPE OF MORTGAGE
RATIO OF INCOME-TO-MORTGAGE PAYMENTS
LOAN AMOUNT

PROPERTY

PROPERTY AGE
NUMBER OF UNITS
APPRAISED VALUE
LOCATION OF THE PROPERTY

OUTPUTS

ACCEPT/REJECT LOAN APPLICANT

RISK ASSESSMENT

Figure 11-5. Inputs and outputs to the neural network in a mortgage loan application.

Image Analysis

Image processing provides a spectrum of applications possibilities (see Figure 11-6). A raw or preprocessed image can be used as an input, with a classification or identification of that image as the network's output. Preprocessing may involve the use of Fourier transforms, specialized functions, known feature extraction algorithms, or other digital image-processing techniques (Duda and Hart 1973). Neural nets that are trained to classify images usually have each output unit represent a different class (see Figure 11-7).

Automated medical testing has been tried for particular medical tests in which the presence or absence of a chemical reaction indicates a positive or negative outcome for the test. The presence of reaction generates a clumpy, nonuniform image from the chemical mixture. The absence of a reaction generates a smooth image from the chemical mixture. This type of classification has been successfully done by a neural network (Figure 11-6). Other medical tests that involve microscopic examination of cells are considerably more difficult to automate. For example, distinguishing normal from abnor-

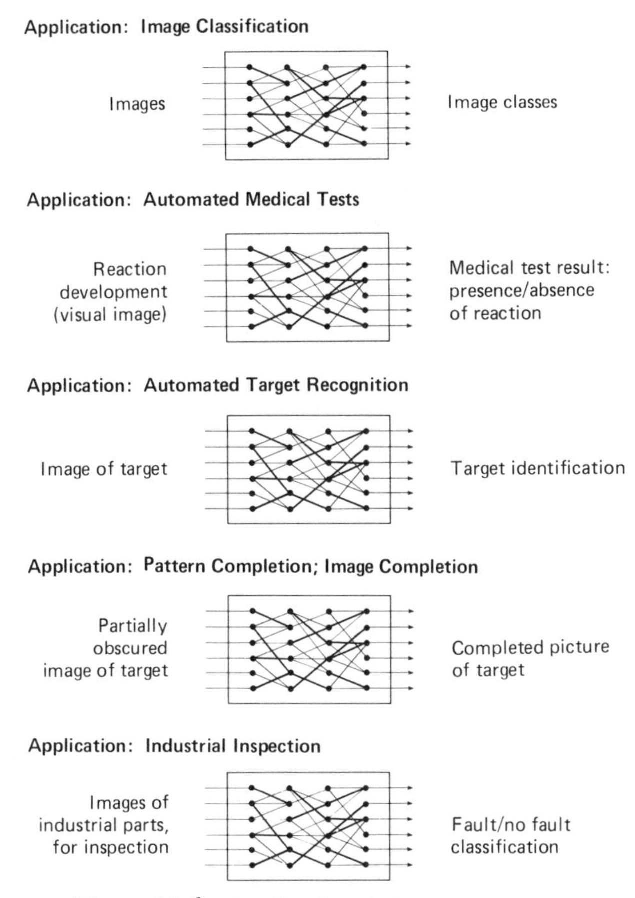

Figure 11-6. Applications in image processing.

mal cells may involve simultaneous use of cell shape, size, clustering, color, and microscope staging information. Problems of this complexity require further work before they will be automated, but appear to be a promising use for neural networks.

Identification of targets is another possible application for neural networks (see Figure 11-6). Here the output units respond to different target images; the targets may or may not be present in a given image. If the target is embedded somewhere in a scene, and must be found, then this is a more complex image-processing problem than classifying a target whose position is known. Figure 11-8 illustrates a candidate target-identification application.

Pattern completion (see Figure 11-9) is the process of filling in missing pieces from a partial pattern. The pattern may be incomplete due to the

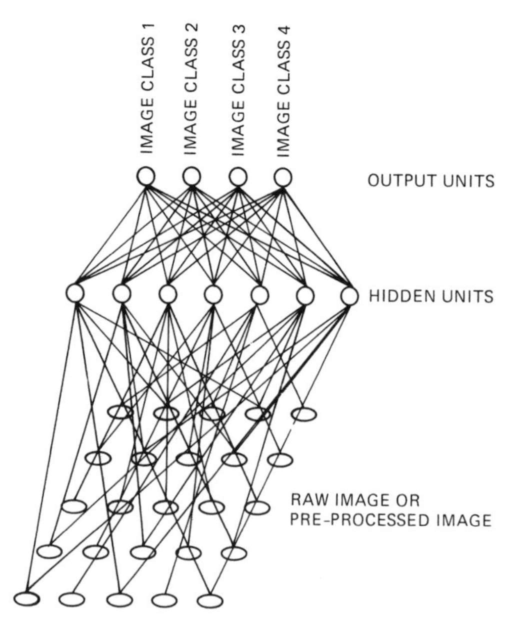

Figure 11-7. A neural network that inputs an image or preprocessed image and produces a classification for the original image.

presence of noise or it may be a pattern that is partially obscured. Examples of two pattern-completion processes are shown in Figure 11-9 (DARPA 1988). The first is a noisy target from the military domain, shown completed by a neural network; the second a partially obscured pattern that is completed when input to the neural net. The network recalls the full pattern.

Typically in a pattern-completion experiment, the neural network is trained on a training set of patterns in which each pattern appears as both input and target output. The network is trained to output the same pattern that is input (homoassociative training). The internal layers of the network develop a distributed encoding of each pattern during training. After training is complete, a partial pattern is presented as input. The network then recalls the full pattern. Typically the output is not perfect; the recalled pattern may itself contain some noise. Optical systems based on neural-like computations have been built to be associative memory systems that can recall a fuzzy form of an entire picture when presented with a partial picture (as in Figure 1-9).

Industrial inspection provides a large domain for candidate neural network applications. Industrial inspection problems (see Figure 11-6) can require a fault/no fault diagnosis, a classification of parts, or an identification of fault type. The classification is often based on an image of a product or component under manufacture.

Figure 11-8. A candidate target identification application. (Adapted from DARPA Neural Network Study 1988.)

In a study by Glover (1988), a variety of visual inspection problems were addressed by neural networks. These included problems in bottle sorting and analysis and analysis of defects in sponges and dials. He compared the Fischer linear discriminant (FLD), back-propagation network (BPN), and counter-propagation network (CPN) methods. Back-propagation usually performed better or comparable to the linear discriminant. In many cases, BPN generalized better than CPN. Figure 11-10a summarizes some of the results from this study, and Figure 11-10b gives a sketch of the bottle types classified for one of the experiments. Preprocessed images were used as inputs to the neural networks.

Diagnosis

Diagnosis is the recognition and identification of the nature or cause of a situation, usually negative. Diagnoses may be made to identify a medical condition, a machine fault, an industrial fault condition, or similar problem. Diagnosis is treated as a pattern-mapping process when a neural network is

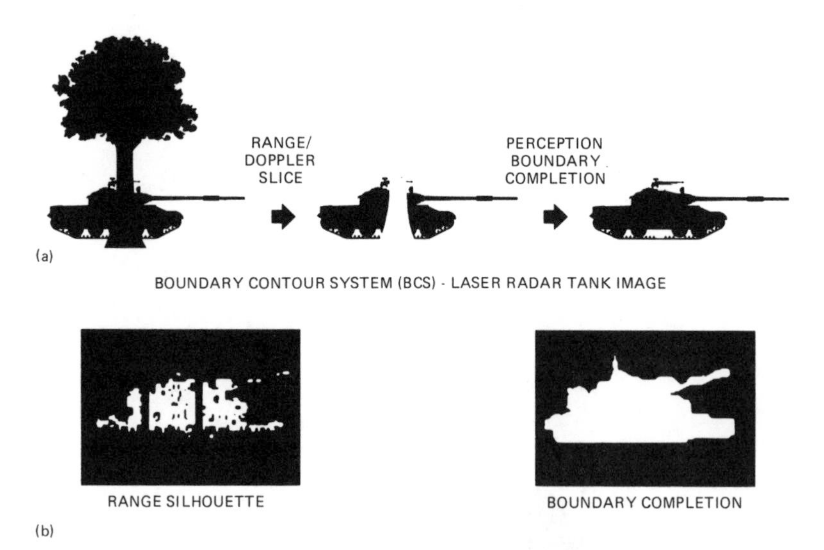

Figure 11-9. Two boundary completions done by neural networks. (a) From a noisy picture. (b) From a partially obscured object. (DARPA Neural Network Study 1988.)

used to perform the diagnosis (see Figure 11-11). The input pattern for the network is the set of data or knowledge about the condition or fault; the output pattern is the diagnosis. The diagnosis may be a specific disease that is identified, a specific engine fault that has been implicated, or a specific cause for the observed problem.

A rule-based approach to automating diagnosis requires a human expert to formulate the rules by which the data can be analyzed. Experience has shown that the set of rules required to accomplish diagnostic tasks is often quite large. Furthermore, developing rules and related heuristics is extremely time consuming because the expert must understand the system's failure modes in order to begin formulating the rules. A rule-based approach does not itself discover the distinguishing features of each fault. In contrast, neural networks offer the potential for rapid and accurate classification of diagnostic patterns. Diagnosis problems utilize a neural network's ability to deal with large data sets, incomplete data, and situations in which the diagnostic rules are not known a priori.

Medical diagnosis can be exemplified by a study in which a dermatology expert system (Figure 11-11) was built with a back-propagating neural network (Yoon et al. 1989). The trained network, DESKNET, is able to diagnose 10 different skin diseases based on a set of 18 symptoms and test results.

EXPERIMENTAL CLASSIFICATION RESULTS

FLD: Fisher linear discriminate; BPN: Back Propagation network; CPN: Counter Propagation network

(Values reported; classification errors/images in testing set)

	TESTING ON		TRAIN AND
APPLICATION EXPERIMENT	TRAINING SET	TEST-ONLY IMAGES	TEST ON COMBINED SET
Soda bottles - 6 classes			
FLD	--	--	--
BPN	10/576	8/576	0/1152
CPN	10/576	6/576	2/1152
Soda bottles - 6 classes			
FLD	--	--	22/1152
BPN	17/576	8/576	1/1152
Soda bottles - 6 classes			
FLD	--	0/864	--
BPN	0/960	0/864	--
CPN	0/960	44/864	--
Soda bottles - 6 classes			
FLD	--	3/864	--
BPN	0/960	7/864	--
CPN	7/960	46/864	--
Syrup bottles - 8 classes			
FLD	--	--	2/1600
BPN	1/1400	1/200	1/1600
CPN	1/1400	1/200	--
Shampoo bottles - 7 classes			
FLD	2/280	9/70	--
BPN	0/280	2/70	--
CPN	4/280	5/70	--

(a)

Figure 11-10. (a) Results from a series of classification problems where three different analysis methods were used on each problem: Fischer linear discriminants (FLD), back propagation (BPN), and counterpropagation (CPN). The analysis was to discriminate between different types of defects.

Figure 11-12 illustrates the network with some of its inputs and outputs. The neural network takes the list of symptoms as input, and outputs its diagnosis by means of activating an output unit that represents a specific disease. The training set consists of a set of symptoms observed in an individual patient, with an expert's diagnosis used as the desired output. A variety of methods for analyzing how the network made its decisions have been suggested for DESK-NET (Yoon et al. 1989).

(b)

Figure 11-10 (cont.). (b) Different categories for the syrup bottle discrimination task.

Machine fault diagnosis and failure analysis is an important commercial application for automated diagnosis systems. Successful applications studies have been done for jet engine diagnosis, rocket engine diagnosis, and automotive engine diagnosis. A study of jet engine diagnosis used a simulated engine to provide hypothetical sensor data for various engine faults (Dietz, Kiech, and Ali 1989). The diagnosis network (Figure 11-13) utilized backpropagation, and was divided into many subnetworks, each with the task of evaluating an individual fault characteristic. Some of the subnetworks evaluated duration or severity of the faults, while others evaluated fault type. Figure 11-13 illustrates a part of the structure of this system. The training set's input patterns consisted of data taken from imaginary sensors in the simulated jet engine. The sensor data usually varied continuously over time, and the network was able to distinguish different types of trends in this data. Desired outputs in the training set were the faults simulated in the jet engine. The network performed with a high degree of accuracy for this system.

A study using neural networks to diagnose faults in automotive control systems (Figure 11-11) has produced a successfully trained network (Marko

Application: Diagnosis

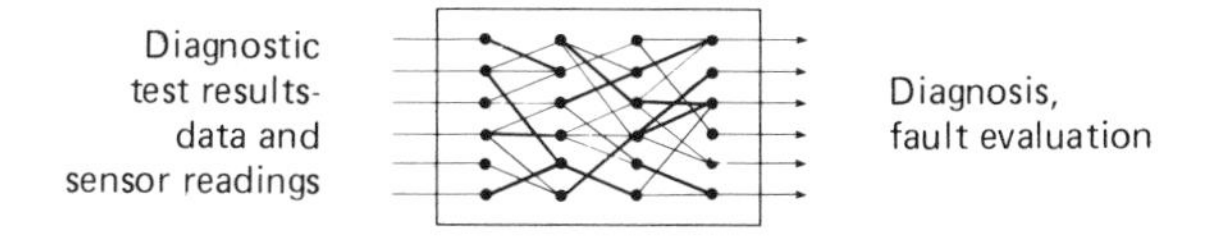

Application: Dermatology Diagnosis

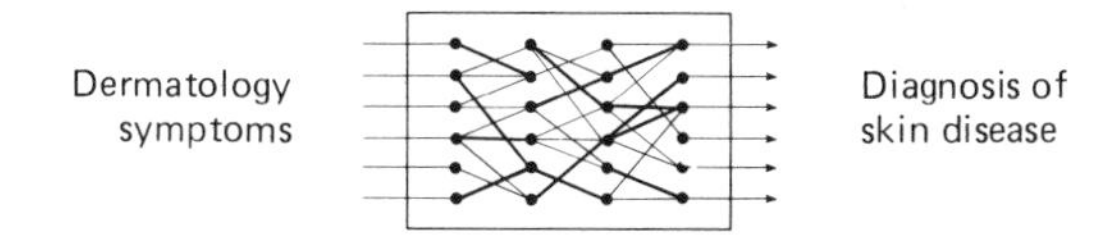

Application: Jet Engine Fault Diagnosis

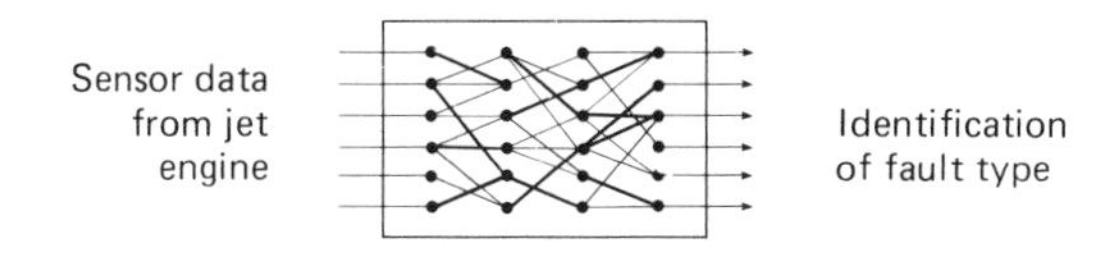

Application: Automotive Control System Diagnostics

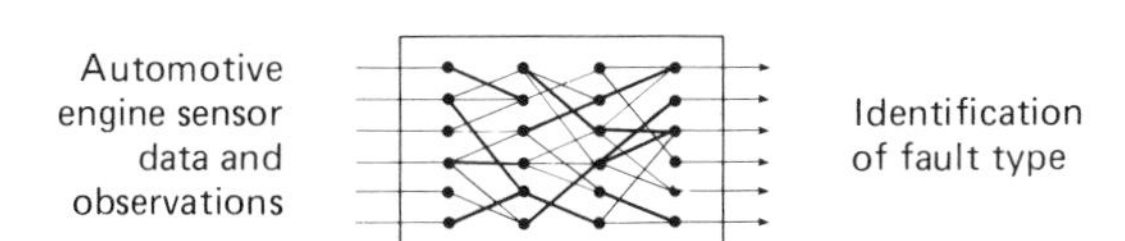

Figure 11-11. Diagnosis applications.

et al. 1989). Researchers applied a variety of neural network paradigms to diagnose engine faults automatically based on data taken from sensors in idling car engines. Their training sets incorporated data from 26 different faults (shorted plug, open plug, fuel injector, broken manifold pressure sensor, etc.) For each fault, 16 data records were taken from engine sensors during engine idling. A comparable test set was also recorded. Results showed that the trained neural networks were able to achieve 100% accuracy for this type of fault diagnosis. The neural networks matched the accuracy of the expert diagnostician and in addition diagnosed the problem faster. This is a significant advance because the problem of diagnosing faults in electronic

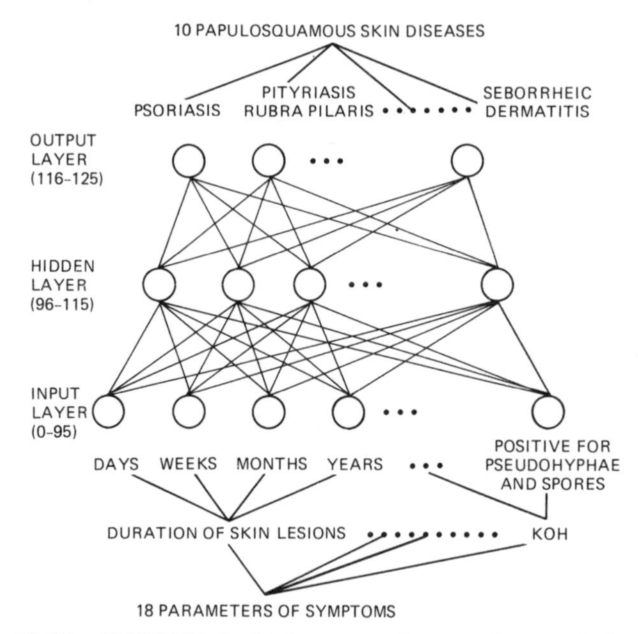

Figure 11-12. DESKNET, the back-propagating neural network that diagnosis skin diseases. (From Yoon, Brobst, Bergstrasser and Peterson. A desktop neural network for dermatology diagnosis. *J. Neural Network Comp.* 1989.)

control systems is a concern due to the growing complexity of the task. Neural networks appear to have the potential to handle the large volumes of information obtained from an automobile's mechanical system.

Automated Control

Neural networks also show promise in control applications. Control problems cover a wide range of complexity, from simple systems such as balancing a broom to complex systems such as autonomous control of a moving vehicle.

The classic control problem that can be solved by a neural network is broom balancing (see Figure 11-14). A broom is inverted on a cart that moves back and forth in an effort to balance the broom on its end. The broom is restricted to a two-dimensional environment (it is not allowed to fall sideways). Successful balancing of a real broom on a real cart was first done by Widrow with an adaline system. A more recent study shows the problem solved in simulation mode — the cart and broom were simulated and controlled by an adaline computer program (Tolat and Widrow 1988). Figure 11-15 shows sample inputs and outputs from this simulation.

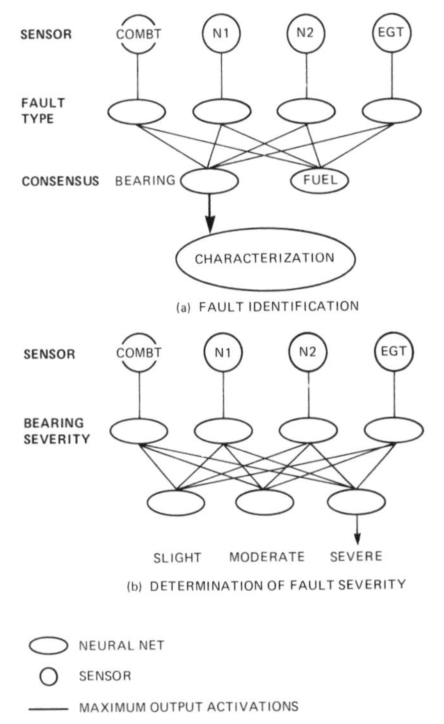

Figure 11-13. Component networks from a back propagating neural net system that diagnoses jet engine faults. (a) Fault identification. (b) Determination of fault severity. (From Dietz, Kiech, and Ali. Jet and rocket engine fault diagnosis in real time. *J. Neural Network Comp.* 1989.)

Robotic control problems are more complex than the classic broom-balancer, and, if solved, would provide a commercially viable application for neural networks. Robotic control studies have focused on the "inverse kinematic problem": given a target position for the end of a robot arm, specify the joint positions and movements of actuators needed to obtain that target position (see Figure 11-14). This is a difficult problem because control of robot manipulators usually involves many actuators that interact with each other, with many degrees of freedom and many ways to carry out a task. Furthermore, they may have to deal with variable loads and a varying or uncertain environment.

There are two major approaches to providing input to the neural network. First, visual sensors are sometimes mounted on the robot arm to view the scene, or the sensors are fixed nearby. The neural network then attempts to map a picture of an object in its visual field to the movements necessary for its arm to reach that object (see Figure 11-14). The second approach simply

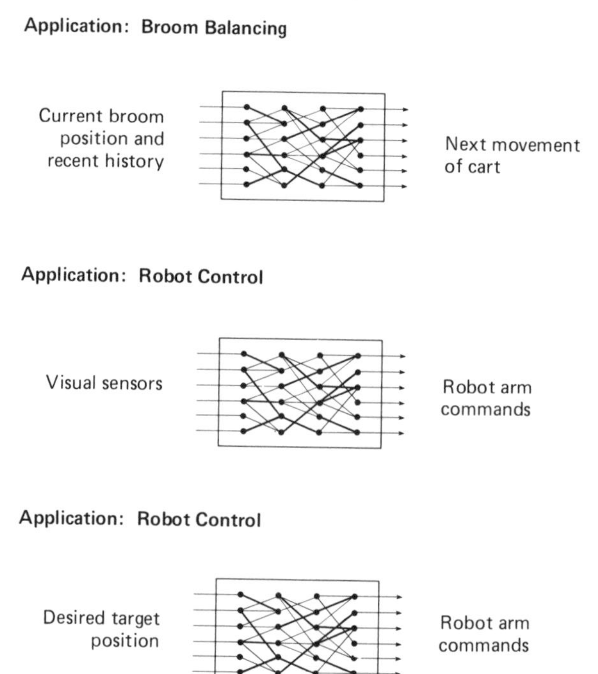

Figure 11-14. Robotics and control applications.

specifies the target position for the robot arm (in three-dimensional coordinates). The neural network takes the target position as input, and as output specifies the actuator control commands required for its arm to get to the target. Training is simple—a human moves the robot arm to the target position. Some success has been obtained with these robot arm studies (Martinetz, Ritter, and Schulten 1989), as illustrated in Figure 11-16.

Speech Analysis and Generation

Speech analysis problems include automated speech recognition and text-to-speech translation. Both problems, if solved, would be useful for a wide variety of commercial applications. A number of speech-recognition tasks have been attempted with neural networks (see Figure 11-17), with success obtained on simplified speech-recognition problems. Current studies have been limited in the following ways: recognition of phonemes or simple words but not more complex words, recognition of only a limited vocabulary, and recognition limited to only a single speaker or a few speakers. Difficult-to-

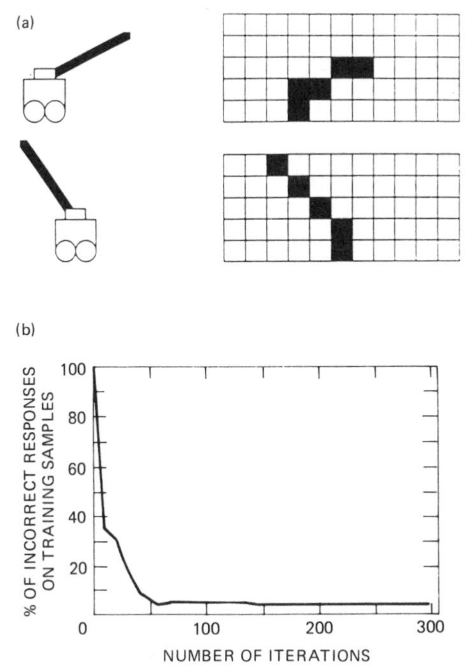

Figure 11-15. (a) Two examples of computer-generated pendulum and cart images. The pictures on the right are 5 × 11 quantized images of the pendulum and cart figures on the left. (b) Learning curve for a one-unit network using two 5 by 11 images as the input. Current and past images were used as input. (From DARPA Neural Network Study 1988.)

master aspects of speech include its temporal nature and the lack of divisions between words in continuous speech. Several studies have had encouraging results (Waibel 1988; Waibel et al. 1989; IJCNN 1989) and the TDNN approach (discussed in Chapter 5) appears quite promising.

The text-to-speech problem is exemplified by NETtalk. For commercial applications we need better performance than was obtained in the original NETtalk investigations. The NETtalk program, however, did learn pronunciation rules faster than comparable text-to-speech systems that do not use neural networks.

More Applications

A preliminary study of sonar signal discrimination has been done for a simple two-category classification problem (Gorman and Sejnowski 1988). A neural network was taught to distinguish the sonar returns (see Figure 11-18) from

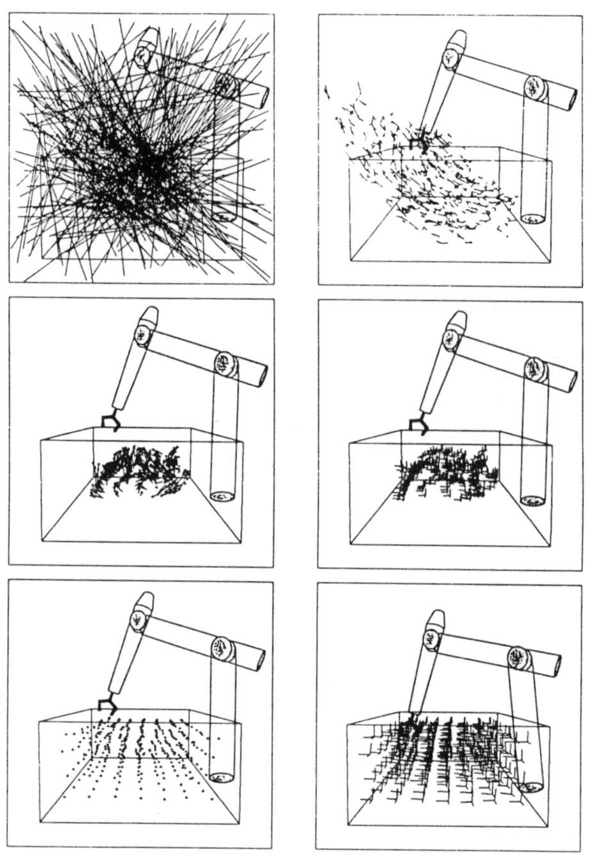

Figure 11-16. The performance of the network at the start (top), after 6,000 iterations (middle) and after 30,000 iterations (bottom). The pictures on the left show the end effector locations (crossmarks) resulting from visual input, together with their deviation (appended line) from the target locations. The right pictures show the reaction of the end effector to small test movements parallel to the borders of the work space. (From Martinez et al. 3 – D neural net for visuomotor coordination of a robot arm. *Proc. IJCNN.* © *1989 IEEE.*)

two different objects: a rock and cylinder. This classification problem is a difficult one for a human to do. Performance of the trained network was close to or better than human performance. A three-layered back-propagating neural network was used, and networks with different size hidden layers were compared. Data was preprocessed to provide input vectors consisting of 60 bins going to 60 input units. The high performance levels of the network showed that the neural network approach holds much promise for solving real sonar recognition problems, which are considerably more complex. Fig-

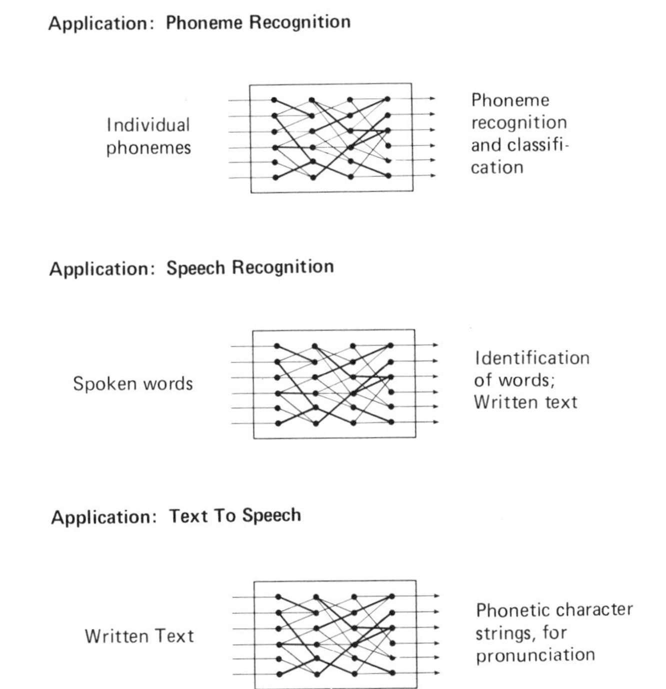

Figure 11-17. Applications for speech processing and generation.

ure 11-19a,b shows the sonar recognition network and also illustrates the preprocessing of the data that was done for this study.

An unusual application for neural networks is in the compact encoding of information. A neural network can actually be trained to provide a compact information coding scheme for a set of patterns (see Figure 11-18). Typically the network has three or more layers, including one particularly small hidden layer. The training set consists of a set of patterns to be encoded, and in each training iteration the network must then learn to compact the information as it propagates through the narrow hidden layer. This type of application can actually give a good encoding scheme that can be used to identify and store only the important information in a series of patterns. The network also reconstructs the original pattern from the encoded version.

Figure 11-20 is taken from a study that compressed images and then reconstructed those images as output (Kuczewski, Myers, and Crawford 1987). In this study, 255 input units and 255 output units were used. There were five layers to the back-propagating network, with only 3 units in the middle layer. The training set consisted of a set of pictures, each used simultaneously as input and target output. The neural net compacted the data in the middle

Application: Sonar Signal Discrimination

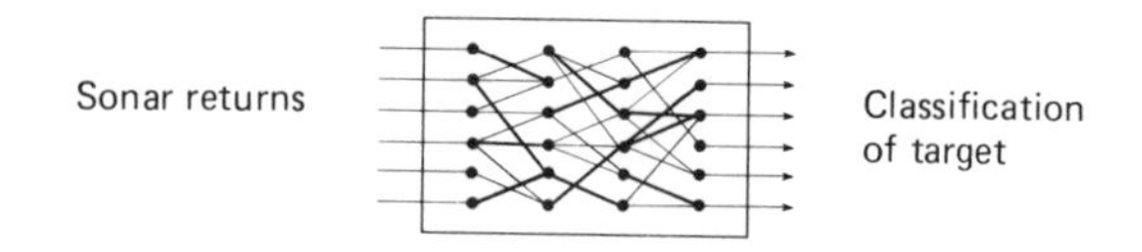

Application: Information Encoding and Compaction

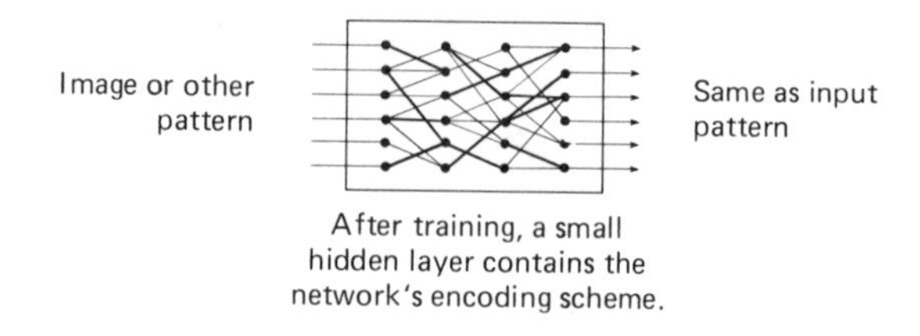

Application: Sequence Recall

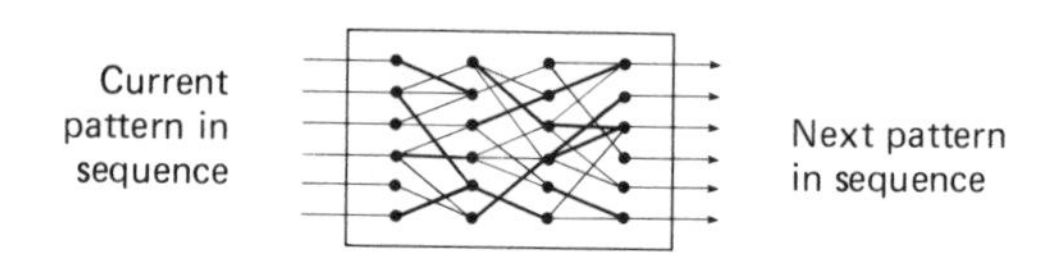

Application: Artificial Critters

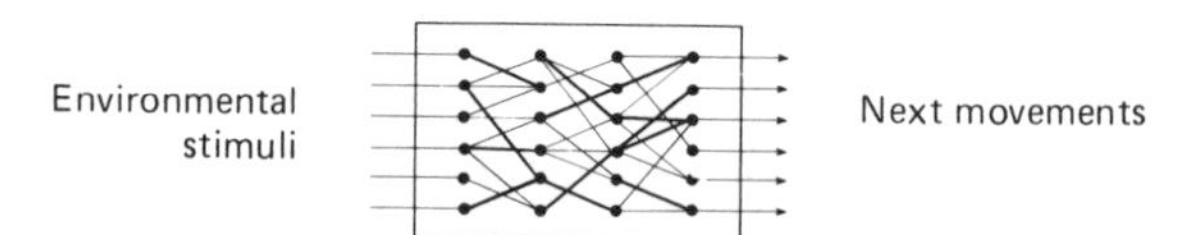

Figure 11-18. More applications for neural networks.

layer and then reconstructed the original image as its output. The resulting picture was quite good. The compact representation that the neural network organized in its middle layer of 3 units was similar to a vector projection from 255 dimensions to only three dimensions. The 3 units in the hidden layer thus retained the important information content of the original images.

Sequence recall can be implemented with a neural network (see Figure 11-18). Each pattern in a sequence is used to recall the next pattern. Thus the training set takes each pattern in the sequence as an input and uses the next

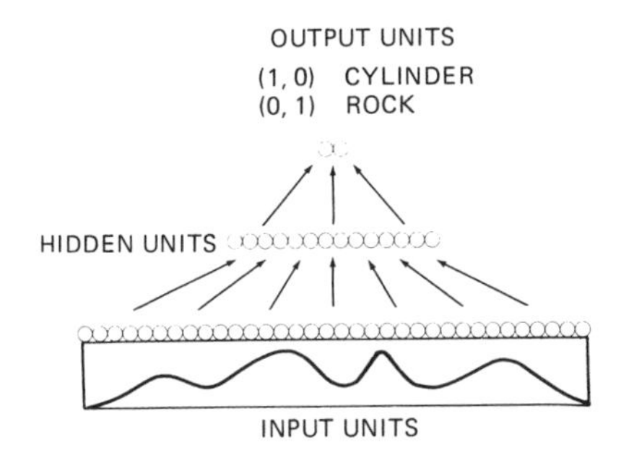

(a)

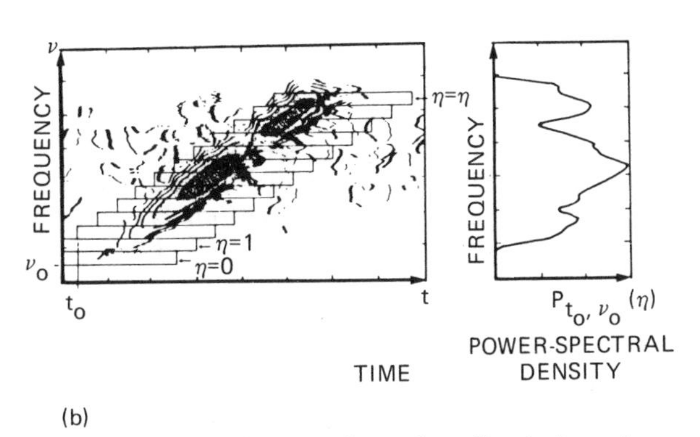

(b)

Figure 11-19. (a) The neural network used to discriminate between sonar returns from a rock and a cylinder. (b) Preprocessing done on the sonar return data. Frequency is plotted versus time on the left. At each frequency, a window is placed over the area with maximum return. The return in that window is summed appropriately to provide a power-spectral density at right. The resulting plot is two-dimensional, and may be input to a neural network with a linear row of input units. (From Gorman and Sejnowski 1988.)

pattern in the sequence as the desired output. In some cases the neural network not only learns the sequence but can recall it from a noisy version of the first pattern, decreasing the amount of noise as each new pattern is recalled. Figure 11-21 shows an example of sequence recall combined with noise reduction, obtained from Kanerva's sparse distributed memory (SDM) scheme. The SDM scheme is similar to the Hopfield network (Keeler 1988) and also shares similarities with the organization of the cerebellum (Kanerva 1988; Kanerva 1989).

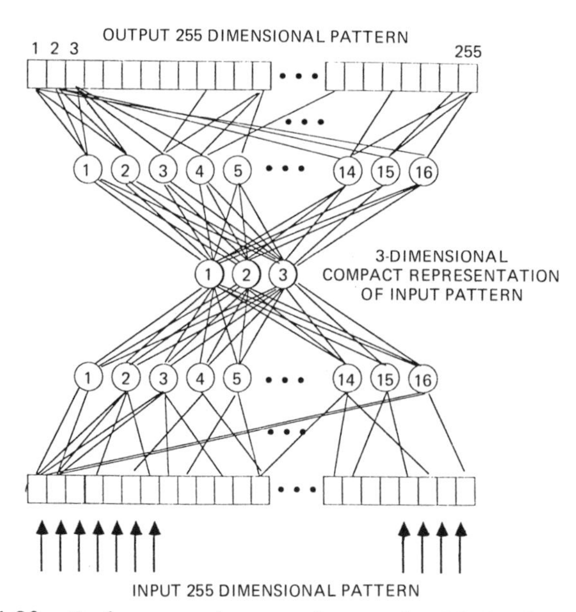

Figure 11-20. Back-propagating neural network with five layers. The input and output layers have 255 units, whereas the middle hidden layer has only three units. (From Kuczewski, Myers, and Crawford. Explanation of backward error propagation as a self-organized structure. *ICNN Proc.* © *1987 IEEE.*)

Perhaps the most playful application has been the construction of "artificial critters" with neural networks (Winter 1989). An "artificial critter" is an adaptive system that moves about an environment and responds to it (see Figure 11-18). One such simulated "critter" was constructed from a neural network that took visual stimuli as input from its visual field and produced as output the next motions of the critter (e.g., right/left/up/down). The network was taught to evade prey that approached in its visual field. Other artificial critters or vehicles have been taught to be attracted to light, to shun light, or to circle about a light source. Some research has been done on teaching particular behavior patterns to artificial critters. Current systems can learn to adapt to their environments but are not yet capable of complex behaviors, which usually require a store of memories.

There is a long path from research to proven applications. Most applications for neural networks are currently under study, with success noted on simplified problems. The difficult part of the studies will be to construct deployable systems with neural networks. If investigators prove that neural networks solve problems that are not solvable by other techniques or show that neural networks make implementations possible that would not otherwise have been feasible, still more work will be required.

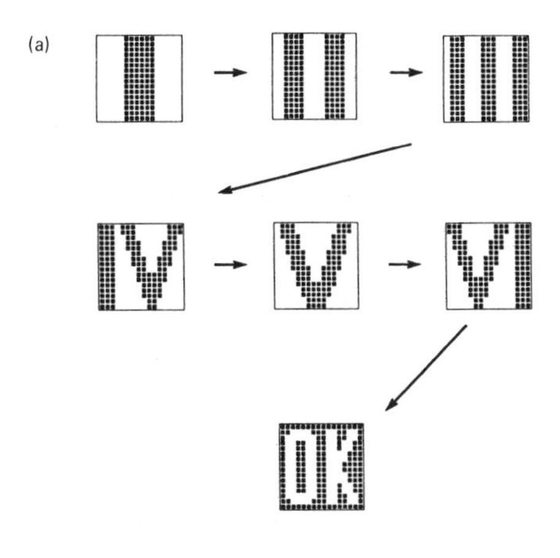

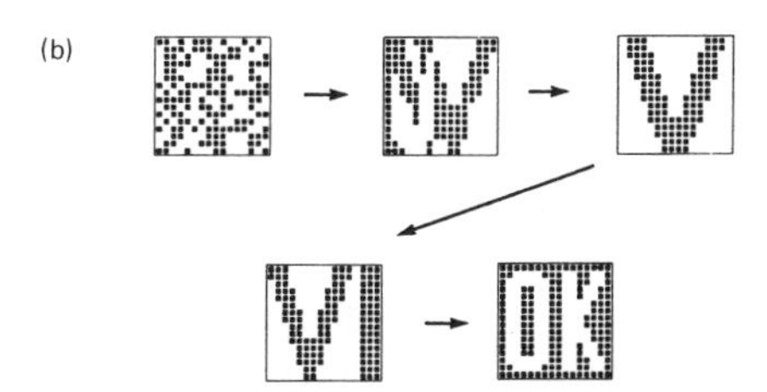

Figure 11-21. (a) A series of seven 256-bit patterns that were written into a sparse distributed memory. $A \rightarrow B$ means the pattern B is recalled when the SDM is presented with pattern A. (b) Reading out a sequence from the SDM, starting with a noisy instance of the Roman numeral III. (From Kanerva 1989. © IEEE.)

A challenge in developing proven applications is obtaining adequate data for a neural network to learn successfully. To build its map successfully, a neural network usually requires a relatively large set of training data, and sometimes simulated data is used to supplement sources of real data. The training data must also be well organized in the sense that discernable patterns must be present before the network can be trained to perform the expected pattern-mapping task. In neural network applications, each candidate application requires its own mapping and the training network organizes a different pattern map for each application system. The training data requirements vary with the applications domain, the type of neural network used, and its size.

BUILDING YOUR OWN APPLICATION

In building a neural network application or in doing a research study, a developer with neural network software can think solely in terms of the neural network and does not have to become involved in sequential programming steps or algorithms. The developer's approach is to build a network using various neural network components and then to allow self-organization to take place during training. The developer can focus exclusively on the structure of the network and the choice of training data and does not have to deal with the details of programming or the ordered nature of algorithms.

Thus, with the proper software, neural network development becomes a creative building job. The builder lays down processing units or layers of processing units on the screen of a computer. Sizes of each layer are chosen. Units are interconnected in layers or individually. The developer may have an automatic command that fully interconnects two layers or may specifically build interconnections, picking out the origin and target unit for each connection.

The developer then has the responsibility of providing training and test data for the network. There are two parts to this job: Data must be assembled for training and testing, and then a representation must be designed for the input and output to the network. In weather forecasting, for example, the training and test sets can be taken from two months of weather data at a particular location. The representation can be to dedicate three units to different temperature ranges, three units to different pressure ranges, and a single unit for the presence or absence of precipitation.

The network's internal computations must be chosen by the developer. The network response to an incoming pattern may be dictated by a chosen paradigm. Alternatively, a developer can specify his or her own summation and transfer functions for the processing units, constructing a variation on one of the standard paradigms. Competitive layers may be selected for inclusion. The adaptation rules also can follow prespecified paradigms or be specially tailored.

Training the network successfully involves many choices and training experiments. First, a training schedule is needed. The training schedule includes the number of training iterations to be done and any changes that are made during training to the learning rate parameters. When a network is trained, a learning curve can be observed. Training sometimes involves many sessions, with different training sessions used to continue training on the same network or to train different networks for comparison. If training experiments are done, the training data and network parameters typically are varied, along with varying the paradigm, network size, and interconnection topology. From these experiments, the developer learns which configurations train the network most successfully for the application at hand. Simulated data may be used to enhance the training set—data may be simulated, or noise may be superimposed on real data to provide the network with more training examples and to accustom it to the noise typically found in realistic applications environments.

The developer is thus an architect for a neural network. With this battery of building blocks and components, the neural network can be built and trained for specific applications. The resulting network is then evaluated for performance. Compared to the sequential programming of simple algorithms, this is a new approach to computing and a new way of thinking.

FUTURE DIRECTIONS AND ISSUES

Computational models with neural network architectures have progressed tremendously since their inception in the 1950s. Advances over four decades have brought us from simple two-layered architectures that required cumbersome hardware implementation to the simulation of hundreds of thousands of processing units on a digital computer. Biological neural models have advanced from simplified models of binary-state neurons to the simulation of networks of neurons with many biological details included. The divergent goals of modeling biology and building better computational tools have provided cross-fertilization for both endeavors.

Specific issues and possible limitations remain to be addressed. Problems such as information representation and scaling up will be key to increasing the power of neural networks. Real-world problems such as vision and speech processing are very complex, and their solution requires synthesizing large amounts of data and integrating the many pieces of information obtained.

Scaling up the size of a neural network is often required in developing an application. Whereas an initial applications study may use tens or hundreds of processing units, the final network may require thousands or tens of thousands of processing units to solve a complex real-world problem. In fact, a DARPA study (1988) has estimated that speech recognition would even require 10^5 connections, and that vision would require up to 10^{10} processing units.

A number of difficulties can arise when scaling up the size of a neural network (Will 1989). First, the computational time can be prohibitive. In addition, a larger and more complex network results in a larger surface on which the network performs a minimization during learning; this in turn can paralyze the network at a local minimum.

One approach to the scaling up of neural networks is to build accelerated hardware, such as customized very-large-scale-integrated (VLSI) circuits. Integrated circuits would undoubtedly provide a great deal of acceleration to neural network computations, and would, in addition, implement the neural network in a very compact form. The ability to implement larger networks in VLSI is ultimately limited, however, due to design constraints such as the maximum density of interconnects. Another approach to scaling up is to build a larger network from many smaller network subunits. This approach has been successful in speech-recognition studies (Waibel et al. 1989). Yet another approach is to develop faster learning algorithms and architectures. An additional issue sometimes remains when scaling up a network—the larger network may not in fact solve more difficult applications problems. For some applications, the network paradigm itself may have to be made more complex before the real-world applications problem can be solved.

Representation and preprocessing can influence the performance level of a neural network and can help solve the problem of addressing the high degree of complexity in real-world problems. Some successful image-classification studies have used preprocessing of images with Fourier transforms and with other image-processing techniques that are sensitive to particular types of features in the data (e.g., line orientations, edges, color, shading, textures, etc). Speech-recognition studies make use of preprocessing techniques such as Fourier transforms, Kalman filters, and other methods before data is entered into the neural network. Even the Xor problem can benefit from a representational change that simply takes the inverse of one of the inputs before presenting it to the neural net. The resulting problem is much easier to learn. Thus, preprocessing of data can perform some of the task of mapping or classifying patterns, leaving the network more likely to solve the remainder of the problem.

Other issues that require further work include assessments of capabilities, accountability, and reliability in trained neural networks. A key challenge is to establish that neural network architectures provide unique capabilities that go beyond other methods such as traditional statistics and optimization routines. If not, then neural network architectures may still be a hardware advance by providing a convenient implementation for a broad spectrum of computations.

Ideally we would like to make a neural network's answers traceable so that the network is "accountable" for its decisions. Several possibilities have already been suggested by researchers (Yoon et al. 1989), although traceability in large neural networks appears difficult or, in some cases, incomprehensible.

A further concern is the reliability of a trained system. Neural networks must be tested for performance after training is complete, and investigators must use appropriate data and experiments for this testing. Testing data should reflect the data to be encountered while the neural network is in use, and testing experiments should use as broad a spectrum of inputs as possible, including inputs that may result from unlikely circumstances. Neural networks do not yet have theoretically proven performance levels; performance must be tested. It should be noted, however, that a performance advantage of neural networks arises from the fact that neural networks continued to perform well in spite of some damage.

Although we have pointed out issues and limitations, the future of neural networks appears promising. Neural networks are likely to address problems with a great deal of complexity, such as speech recognition and visual processing. These problems require a very sophisticated set of analysis tools. Neural networks are likely to have an impact on an extremely broad base of applications areas, including financial analysis; image processing in defense, medical, and industrial domains; diagnosis in medical and commercial domains; robotic control; speech recognition and synthesis; sensor data classification; and information encoding. Neural networks appear to be good at solving problems such as pattern recognition, pattern mapping, the analysis of noisy patterns, associative lookups, and pattern completions, in addition to providing systems that can learn and adapt during use. Successful applications have been designed, built, and commercialized, and investigators continue to extend this success.

In tandem with the attempt to build better artificial systems, neural network modeling will also maintain a place in neuroscience. Because biological systems are so complicated in structure and physiology, it is valuable to have simulation models that help predict their dynamic activity. Most simulation models today incorporate only a subset of the structure found in biology. Artificial neural network models contain only a few of the most basic biological structures, and often contain additional structures that are not found in biology. Even in biologically oriented simulation models, we are far from simulating an actual biological system with its complex microstructure. Simulation studies of biological neural models, however, will contribute to our understanding of the brain as well as provide input to builders of artificial neural systems.

Although evolution built the first neurobiological system, humankind's relentless search to emulate Nature has motivated the neural network approach to computing. As the capabilities of neural network architectures become understood more completely, both their limits and abilities provide lessons for computer engineering and brain research. The continuing effort to evolve human-made systems that mimic parts of biological architectures has, with the advent of neural networks, taken a giant step forward. The success of biological neural systems is certain and self-evident. The performance and utility of artificial systems is gradually unfolding as research progresses, and will continue to have an impact on our lives.

References

Collins, E., Ghosh, S., and Scofield, C. 1988. An application of a multiple neural network learning system to emulation of mortgage underwriting judgments. *IEEE International Conference on Neural Networks, II*:459–466.

DARPA Neural Network Study. 1988. Fairfax, Virginia: AFCEA Press.

Dietz, W. E., Kiech, E. L., and Ali, M. 1989. Jet and rocket engine fault diagnosis in real time. *J. Neural Network Computing* 1(1):5–18.

Duda, R. and Hart, P. 1973. *Pattern Classification and Scene Analysis*. New York: Wiley Interscience.

Glover, D. 1988. A hybrid optical fourier/electronic neurocomputer machine vision inspection system. *Proceedings of the Vision 1988 Conference*. Dearborn, Michigan: Society of Manufacturing Engineers.

Gorman, R. P. and Sejnowski, T. J. 1988. Analysis of hidden units in a layered network trained to classify sonar targets. *Neural Networks* 1(1):75–89.

International Joint Conference on Neural Networks (IJCNN) Vol. I and II. 1989. Piscataway, N. J.: IEEE Press.

Kanerva, P. 1989. A cerebellar model: Associative memory as a generalized random-access memory. *Proceedings of the 34th Computer Society International Conference (COMPCON), Feb. 27–Mar. 3, 1989.*

Kanerva, P. 1988. *Sparse Distributed Memory*. Cambridge, Massachusetts: MIT Press.

Keeler, J. D. 1988. A comparison between Kanerva's SCM and Hopfield-type neural networks. *Cognitive Science* 12:299–329.

Kuczewski, R. M., Myers, M. H., and Crawford, W. J. 1987. Exploration of backward error propagation as a self-organized structure. *Proceedings IEEE First International Conference on Neural Networks* II:89–96.

Marko, K., James, J., Dosdall, J. and Murphy, J. 1989. Automotive control system diagnostics using neural nets for rapid pattern classification of large data sets. *Proc. of International Joint Conference on Neural Networks, II*:13–15.

Martinetz, T. M., Ritter, H. J., and Schulten, K. J. 1989. 3D neural net for learning visuomotor coordination of a robot arm. *Proceedings International Joint Conference on Neural Networks, II*:351–356.

Shea, P. M. and Lin, V. 1989. Detection of explosives in checked airline baggage using an artificial neural system. *Proc. International Joint Conference on Neural Networks,* II:31–34.

Tolat, V. V. and Widrow, G. 1988. An adaptive "broom balancer" with visual inputs. *IEEE International Conference on Neural Networks,* II:641–647.

Waibel, A. 1988. Consonant recognition by modular construction of large phonemic time-delay neural networks. In *Neural Information Processing Systems,* San Mateo, California: Kaufman Publishers, pp. 215–223.

Waibel, A., Hanazawa, T., Hinton, G., Schikano, K., and Lang, K. 1989. Phoneme recognition using time-delay neural networks. *IEEE Trans. Acoustics, Speech, and Signal Processing (ASSP),* 37 (March).

Widrow, B. and Stearns, S. D. 1985. *Adaptive signal processing*. Englewood Cliffs, N.J.: Prentice-Hall.

Will, C. 1989. Modular architectures can help in scaling up neural networks. *Synapse Connection* **3**.

Winter, C. L. 1989. Robotics applications. *Journal of Neural Network Computing* 1(1):66–72.

Yoon, Y., Brobst, R. W., Bergstresser, P. R., and Peterson, L. 1989. A desktop neural network for dermatology diagnosis. *J. Neural Network Computing* 1(1):43–52.

Glossary of Biological Terms

acetylcholine (ACh)—$C_7H_{17}NO_3$, a neurotransmitter.
action potential—a wave of depolarization that typically propagates down an axon, away from the cell body, to the terminal ends of the axon.
afferent—incoming; said of impulses or pathways that carry impulses toward the central nervous system.
amacrine cell—a retina cell with short dendrites that provides lateral interconnections.
amygdala—a nucleus in the limbic system, functionally involved with emotions and behavioral motivation.
anatomical—having to do with the physical structure of a living thing.
arborization—formation of a branched, treelike structure.
auditory cortex—a lateral (side) area of the cerebral cortex, responsible for hearing processing.
axon—a branching fiber that originates at the neuron cell body and functions as the principal output structure of the neuron.
basilar membrane—a membrane wound within the cochlea that vibrates from sound energy and stimulates receptor cells as it moves.
basket cell—a type of cell found in the cerebellum that wraps its axons about Purkinje cell bodies.
bilateral—having to do with both sides.
bipolar cell—a type of cell found in the retina, that transfers information from the receptor cells to the ganglion cells.
brain stem—the stemlike lower portion of the brain, which connects the cerebral hemispheres with the spinal cord.
central nervous system—that part of a nervous system that forms a distinct principal concentration of cords, ganglia, and nerve clusters. In humans, the brain and the spinal cord.

cerebellum—a portion of the brain behind and below the cerebral cortex, responsible for motor balance and coordination.
cerebral cortex—the outer portion of the brain, located above the cerebellum.
climbing fiber—a type of branching axonal fiber that occurs in the cerebellum; its distinguishing characteristic is that it grows to match the dendritic tree of a recipient neuron.
cochlea—a small organ in the inner ear that transforms sound energy to coded nerve signals.
conformational change—a change in the three-dimensional configuration of a biological macromolecule. Typically proteins undergo conformational changes.
convolutions—uneven bends and crevices on the surface of the cerebral cortex.
cortical column—a cylindrical shaped area on the cerebral cortex that is perpendicular to the surface and contains neurons that are dedicated to the same type of stimuli.
cytoplasm—the fluid inside a cell.
dendritic spine—a protrusion on a dendrite that typically synapses to one or more axons, to receive incoming signals.
dendrites—a set of branching fibers located on each neuron that function as a receiving area for incoming signals.
depolarization—a process whereby a cell membrane becomes less polarized, resulting in a smaller difference in electrical potential between the two sides of the membrane.
echolocation—the process of locating objects by sending out an acoustic signal and listening for the characteristics of the echoed response.
efferent—outgoing; said of impulses or pathways that carry impulses away from the central nervous system.
excitatory postsynaptic potential (EPSP)—a postsynaptic response that consists of depolarization and thus serves to excite the target cell towards firing an impulse.
exocytosis—the process whereby the membrane of a synaptic vesicle merges with the cell's membrane and the vesicle then releases its contents into the extracellular space.
fluid mosaic membrane model—a model of membrane structure in which larger molecules are embedded in a phospholipid bilayer. The embedded molecules drift about slowly.
fovea—the central area of the retina, which is the most densely enervated.
GABA—gamma-aminobutyric acid, an amino acid neurotransmitter.
gap junction—a primitive functional connection between nerve cells.

membrane gate—an intrinsic membrane protein that can sometimes provide a pathway for ions to travel from one side of a membrane to the other.

glial cell—a class of cells that are not neurons but nevertheless appear in the nervous system and play a supportive role to the neurons.

glomerulus—a cluster of synapses.

glutamate—an amino acid that functions as a neurotransmitter.

hair cell—a type of receptor cell with one or more hairs that are stimulated. Found in the auditory system (to detect sound) and in the vestibular system (to detect motion).

hippocampus—a structure in the limbic system (with the shape of a curved, elongated ridge) that plays a role in short-term memory.

hyperpolarization—a process whereby a cell membrane becomes more polarized, resulting in a greater difference in electrical potential between the two sides of the membrane.

inferior colliculus—a nucleus in the auditory system.

inhibitory postsynaptic potential (IPSP)—a postsynaptic response that produces hyperpolarization and thus serves to inhibit the target cell from firing.

intrinsic membrane protein—a protein embedded in a membrane.

ion—an atom or group of atoms that carries a positive or negative electric charge as a result of having lost or gained one more electron.

lateral—referring to the side of an anatomical entity.

lateral geniculate—a structure deep inside the brain responsible for preliminary processing of visual signals.

limbic system—a system located deep within the central part of the brain, that includes the amygdala, the hippocampus, and the limbic cortex, and appears similar to primitive brains such as reptilian brains.

lock-and-key mechanism—a mechanism whereby one molecule (usually a macromolecule) binds selectively only to specific molecules, similar to the way that a specific key fits a particular lock.

macromolecules—large molecules, typically found in living things.

medial geniculate—a nucleus in the auditory system.

membrane potential—the difference in electrical potential between the inside and the outside of a cell membrane.

membrane pump—an intrinsic membrane protein that causes ions to move from one side of the membrane to the other. The movement expends energy.

motor neuron—a nerve cell that sends its axon to a muscle and stimulates muscle cells when axon potentials arrive.

muscle stretch receptor—a type of receptor cell, located in a muscle, that sends signals from the muscle back to the central nervous system.

myelin sheath—a thin covering that is wrapped around some axons. It consists of glial cells, and provides insulation to the axon.

neuron—a biological nerve cell.

neurotransmitter—a chemical substance that is released at a synapse by the presynaptic cell, typically when an action potential arrives. The neurotransmitter is received by the postsynaptic cell membrane.

node of Ranvier—A gap in the myelin sheath, exposing the axon at periodic intervals along the fiber, formed by the space between segments of the sheath.

nucleus—a globular structure, typically within the brain, that consists of a dense cluster of nerve cell bodies and synapses.

occipital lobe—the back portion of the cerebral cortex, responsible for visual processing.

olfaction—the sense of smell.

optic nerve—the nerve that goes from the retina to the brain.

Pacinian corpuscle—a type of receptor cell that responds to mechanical movements.

peptide—a short protein.

phospholipid bilayer—a double layer of phospholipid molecules, which forms the basic structure of a cell membrane.

phospholipid molecule—a molecule with a phosphoric acid head and a glyceride tail. The phosphoric acid head has an affinity for water, where as the glyceride tail tends to repel water.

physiological—having to do with the dynamic operations and mechanisms that take place in living things.

project—to send a nerve pathway from one location to another.

Purkinje cell—a type of cell, found in the cerebellum, that has a large dendritic arborization.

receptor cell—a cell that receives sensory stimuli and generates responses to the stimuli. The responses are usually a change in membrane potential and/or an action potential.

receptor protein—a protein that binds, or receives, another molecule, and brings about changes as a result of that binding. Receptor proteins are located on postsynaptic membranes to receive neurotransmitter molecules.

refractory period—the period of time just after a nerve cell has generated an action potential, during which the cell is incapable of generating another action potential. Refractory periods typically last 3–5 ms.

retinal ganglion cell—a nerve cell with its cell body in the retina and its axon going up the optic nerve to the brain.

selective binding—the ability of a molecule to bind exclusively to only one other molecule or to only a few other molecules.

superior olive—a nucleus in the auditory pathway.

synapse—a functional connection between distinct neurons accomplished by near contact of their membranes.

synaptic cleft—the gap between the presynaptic membrane and the post-synaptic membrane.

synaptic vesicles—small vesicles that contain neurotransmitter and are found at synapses, in the presynaptic cell.

teleost—a type of fish with a bony skeleton rather than a cartilage skeleton.

temporal lobe—the lateral lobe of the cerebral cortex.

terminal bouton—a small bulge that is often found at the end of a nerve axon and that synapses to other neurons.

thalamus—a group of nuclei deep within the brain, which serves as a relay station for the major sensory systems that project to the cerebral cortex.

tonotopic map—a topological map that maps audio frequencies across a particular area on the surface of the brain.

topological map—any region of the brain in which the activity of individual neurons corresponds to sensory or motor phenomena and the neurons together form a map of the sensory or motor system.

transducer—a mechanism or structure that changes one form of energy into another (e.g., incoming light is transformed into electrochemical changes in the retina).

Index